Intermediate 2

BIOLOGY

MULTIPLE CHOICE TESTS

Team Co-ordinator
James Torrance

Writing Team
James Torrance
James Fullarton
Clare Marsh
James Simms
Caroline Stever

Diagrams by James Torrance

Hodder & Stoughton

A MEMBER OF THE HODDER HEADLINE GROUP

Orders: please contact Bookpoint Ltd, 130 Milton Park, Abingdon, Oxon OX14 4SB. Telephone: (44) 01235 827720, Fax: (44) 01235 400454. Lines are open from 9.00–6.00, Monday to Saturday, with a 24 hour message answering service. Email address: orders@bookpoint.co.uk

British Library Cataloguing in Publication Data
A catalogue entry for this title is available from The British Library

ISBN 0 340 77534 3

First published 2000
Impression number 10 9 8 7 6 5 4 3 2
Year 2006 2005 2004 2003 2002 2001

Illustrated by James Torrance
Cover photo from Bruce Coleman Collection/William S. Paton.
Typeset by Fakenham Photosetting Ltd, Fakenham, Norfolk.
Printed in Great Britain for Hodder & Stoughton Educational, a division of Hodder Headline Plc, 338 Euston Road, London NW1 3BH by J. W. Arrowsmith Ltd, Bristol.

Contents

Preface

This book has been written specifically to complement the text book *Intermediate 2 Biology*. It is intended to act as a valuable resource to pupils and teachers by providing a comprehensive bank of multiple choice items, the content of which adheres closely to the SQA Higher Still syllabus for Intermediate 2 Biology (to be examined in and after 2000).

Each chapter consists of two tests and corresponds to part of a syllabus sub-topic. The questions vary in type with many testing *knowledge* and *understanding*, some testing *problem-solving* skills and others testing *practical abilities*.

The tests allow pupils to practise extensively in preparation for the examination. The book concludes with two 25-item specimen examinations in the style of the multiple choice section of the externally assessed examination paper.

1 Structure and function of cells

Test 1

Items 1, 2 and 3 refer to the accompanying diagram of a microscope.

1 Which labelled structure is the nosepiece?

A P **B** Q **C** R **D** T

2 The nosepiece should be rotated to click the objective lens into place if the image is

A out of focus.
B poorly illuminated.
C half light and half dark.
D surrounded by air bubbles.

3 If P contains a lens with a magnification of ×15 and S and T give a magnification of ×10 and ×40, respectively, then the levels of magnification possible for this microscope are

A ×15 and ×50. **B** ×25 and ×55.
C ×150 and ×400. **D** ×150 and ×600.

4 Which of the following is the basic unit of life (i.e. the smallest unit that can lead an independent existence)?

A chloroplast **B** molecule **C** nucleus **D** cell

Items 5, 6 and 7 refer to the accompanying diagram of a cell from a leaf of a green plant.

Items 5 and 6 refer to the following possible answers.

A vacuole **B** cell wall
C cytoplasm **D** cell membrane

5 What name is given to structure 1?

6 What name is given to structure 4?

7 Which parts of this plant cell are also present in animal cells?

A 1, 2 and 5 **B** 2, 4 and 5
C 2, 4 and 6 **D** 3, 4 and 6

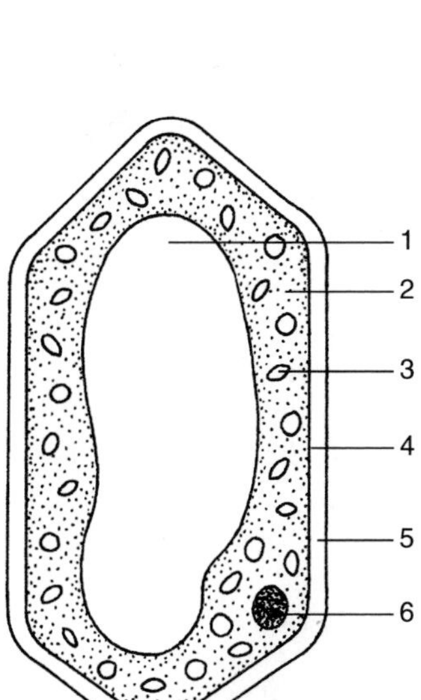

8 Yeast is a

A single-celled fungus.
B multicellular fungus.
C single-celled bacterium.
D multicellular bacterium.

Questions 9 and 10 refer to the following possible answers.

A glucose → CO_2 + water + energy
B glucose → CO_2 + alcohol + energy
C glucose + oxygen → CO_2 + water + energy
D glucose + oxygen → CO_2 + alcohol + energy

9 Which answer represents the equation of aerobic respiration in yeast cells?

10 Which answer represents the equation of anaerobic respiration in yeast cells?

11 The experiment shown in the accompanying diagram was set up to investigate the action of live yeast cells on glucose in the absence of oxygen.

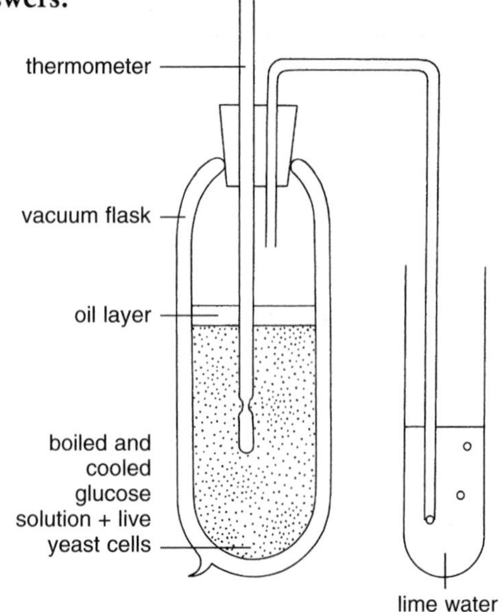

A control experiment should also have been set up where the

A oil was replaced by alcohol.
B live yeast was replaced by dead yeast.
C lime water was replaced by bicarbonate indicator.
D glucose was replaced by fresh fruit juice.

12 In an investigation into wine-making, four test tubes were set up as indicated in the following table.

Condition	Test tube			
	1	2	3	4
boiled and cooled fruit juice	✔	✔	✘	✔
live yeast	✔	✘	✔	✔
boiled yeast	✘	✔	✘	✘
temperature at 40 °C	✔	✘	✘	✘
temperature at 30°C	✘	✔	✔	✔

✔ = present ✘ = absent

Alcohol will be formed in

A both tubes 1 and 2.
B both tubes 2 and 3.
C tube 3 more quickly than tube 1.
D tube 4 more quickly than tube 1.

13 Which line in the following table is INCORRECT?

	Fossil fuel	**Alternative fuel**
A	renewable	non-renewable
B	produces toxic gases following complete combustion	does not produce toxic gases following complete combustion
C	limited resource	unlimited resource
D	coal is an example	alcohol is an example

14 The accompanying diagram shows the results of an experiment where discs of antibiotics X and Y were placed on two Petri dishes each inoculated with a different species of bacterium and incubated at 30 °C for two days.

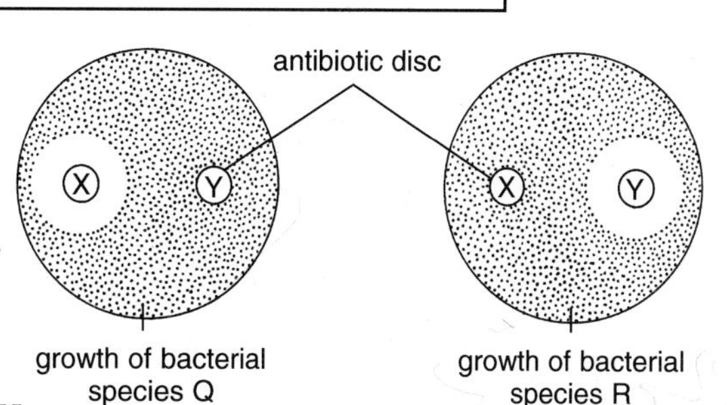

Which of the following statements is CORRECT?

A Bacterial species Q is sensitive to antibiotic Y.
B Bacterial species R is resistant to antibiotic X.
C Bacterial species Q is sensitive to both antibiotics X and Y.
D Bacterial species R is resistant to both antibiotics X and Y.

Items 15 and 16 refer to the accompanying bar chart of a country's production of ethanol using biotechnological techniques.

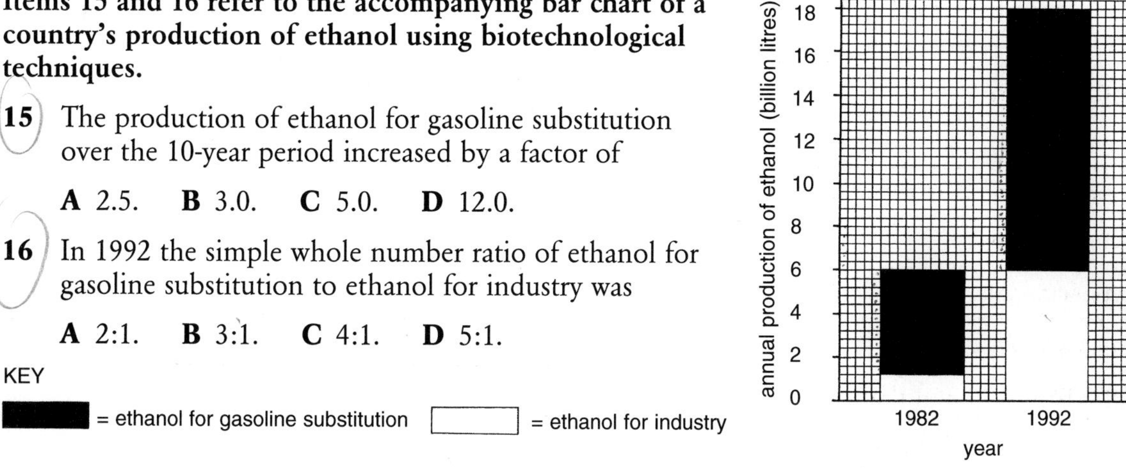

15 The production of ethanol for gasoline substitution over the 10-year period increased by a factor of

A 2.5. **B** 3.0. **C** 5.0. **D** 12.0.

16 In 1992 the simple whole number ratio of ethanol for gasoline substitution to ethanol for industry was

A 2:1. **B** 3:1. **C** 4:1. **D** 5:1.

Items 17, 18 and 19 refer to the following information. Each of the dishes in the experiment was streaked with five different species of bacteria (1, 2, 3, 4 and 5). A strip containing one of four different antibiotics (P, Q, R and S) was then placed across each dish. The diagram shows the results after incubation for three days at 37 °C.

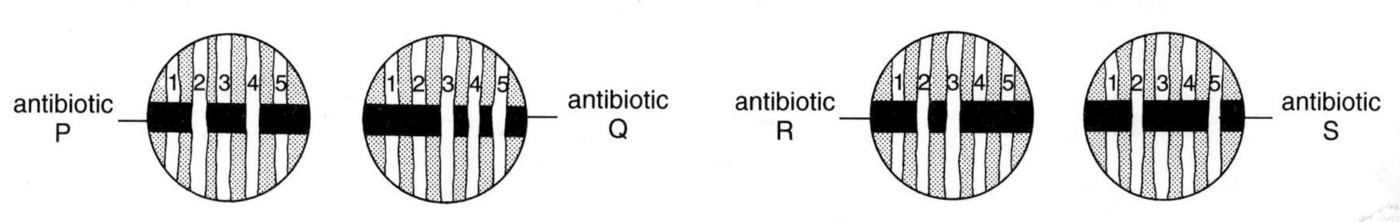

17 Which antibiotic was LEAST effective at preventing bacterial growth?

A P **B** Q **C** R **D** S

18 Which species of bacterium was sensitive to ALL four antibiotics?

A 1 **B** 2 **C** 3 **D** 4

19 Which species of bacterium was MOST resistant to the antibiotics?

A 1 **B** 2 **C** 4 **D** 5

Items 20, 21 and 22 refer to the following information. The accompanying graph shows the results of culturing the fungus *Penicillium* in a large fermenter to produce the antibiotic penicillin.

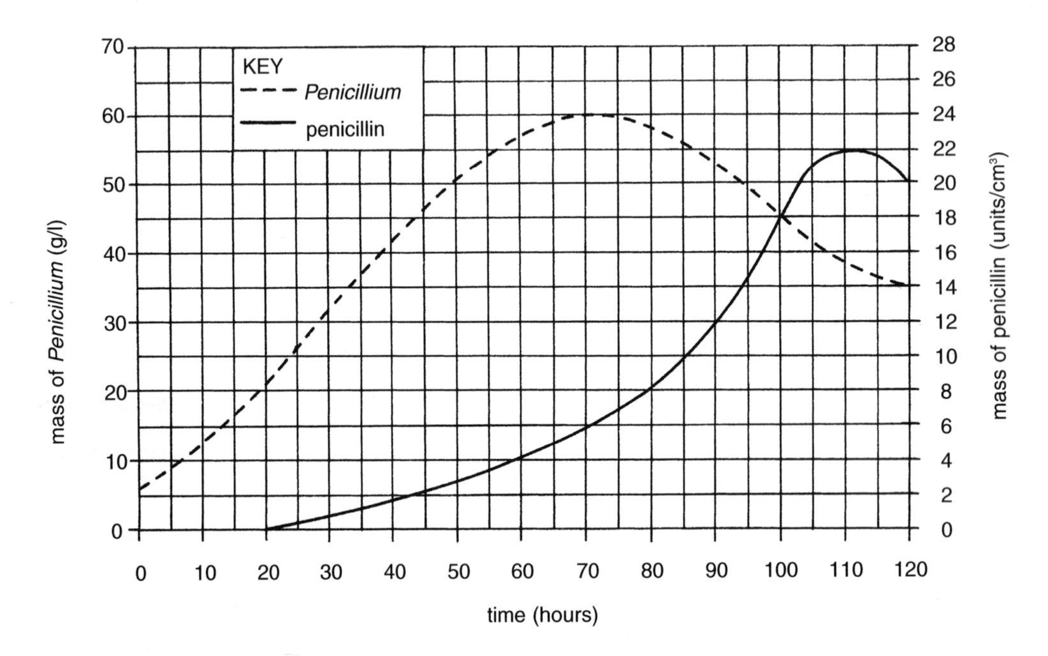

20 The mass of penicillin shows the greatest gain between hours

A 20 and 40. **B** 40 and 60. **C** 60 and 80. **D** 80 and 100.

21 Which line in the following table correctly refers to the mass of fungus and the mass of antibiotic present at 94 hours?

	Mass of *Penicillium* (g/l)	Mass of penicillin (units/cm³)
A	35	14
B	50	14
C	35	50
D	50	35

22 The best time to harvest the fungus and extract the antibiotic would be at

A 70 hours. **B** 100 hours. **C** 110 hours. **D** 120 hours.

23 Which label in the following diagram indicates a way in which a bacterium resists an antibiotic?

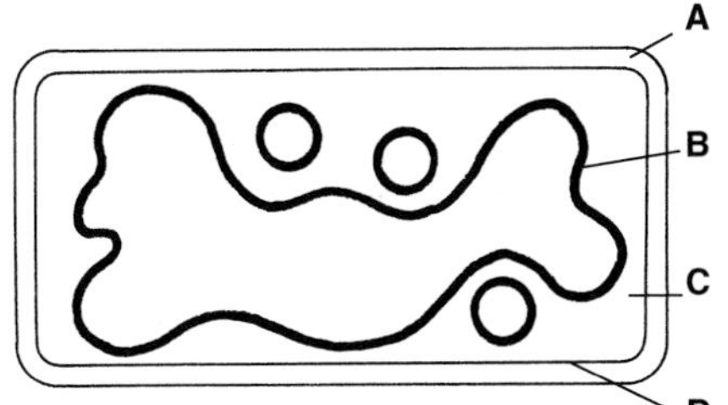

Items 24 and 25 refer to the accompanying diagram which shows an experiment set up to investigate the action of yoghurt bacteria on milk.

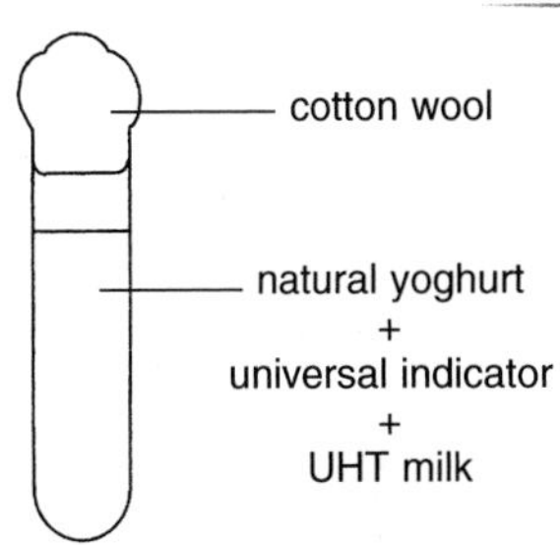

24 Which of the following would act as a suitable control for this experiment?

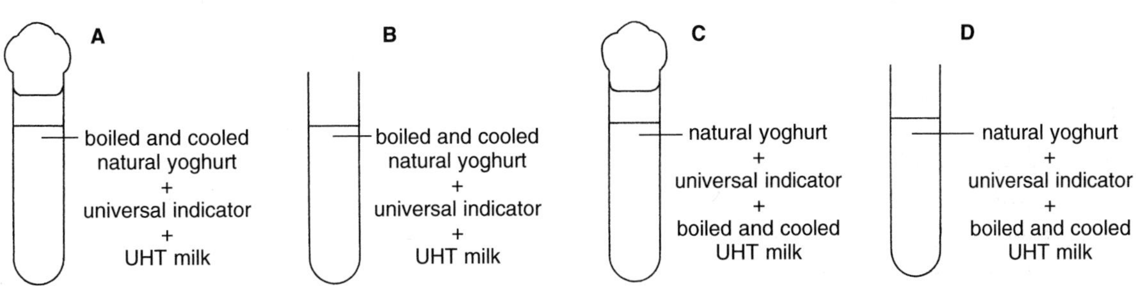

25 Imagine that the experiment and a suitable control have been set up and left for 24 hours in a warm incubator. Which of the following bar graphs correctly represents the results that would be obtained?

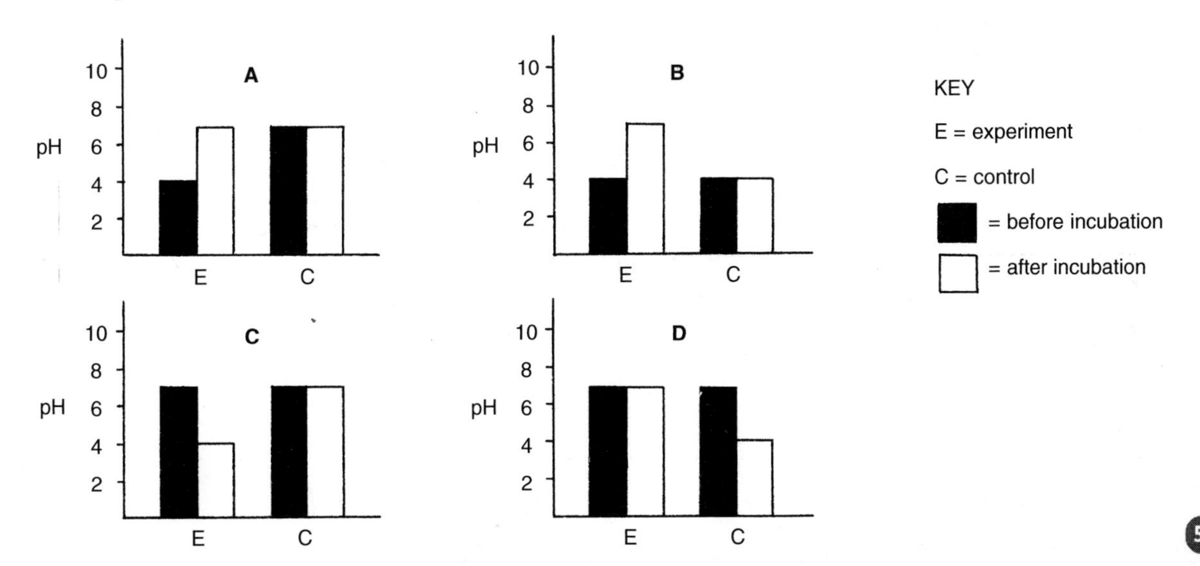

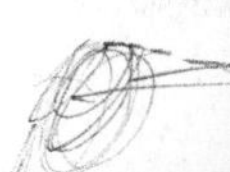

Test 2

Items 1, 2, 3, 4, 5 and 6 refer to the following diagram of a leaf cell.

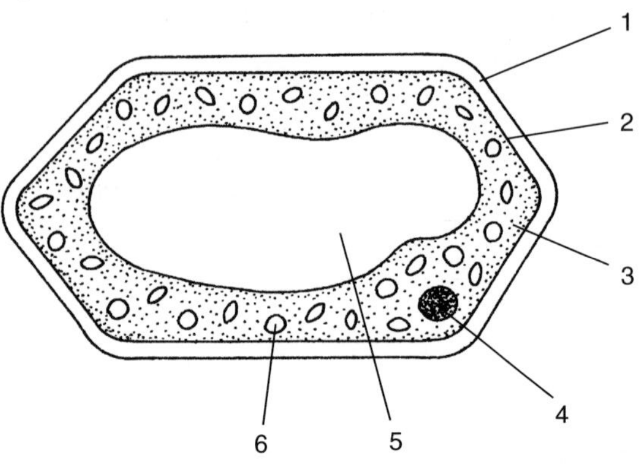

1 A typical animal cell would NOT have

A 1, 3 and 5. **B** 1, 4 and 6.
C 1, 5 and 6. **D** 2, 5 and 6.

2 The structure composed of colourless, jelly-like material which is the site of many biochemical reactions is

A 1. **B** 3. **C** 4. **D** 6.

3 Which numbered structure would yeast lack?

A 1 **B** 2 **C** 4 **D** 6

4 Which structure is composed of cellulose fibres?

A 1 **B** 2 **C** 4 **D** 6

5 Which numbered structure passes information on to daughter cells following mitosis?

A 2 **B** 3 **C** 4 **D** 6

6 The function of structure 6 is to

A control the passage of substances into the cell.
B absorb light energy for photosynthesis.
C store water and solutes as cell sap.
D control the cell's activities.

7 The diameter of a red blood cell is 7 micrometres. Which of the following correctly expresses this measurement as a fraction of a millimetre (mm)? (Note: 1 millimetre = 1000 micrometres).

A 0.7 mm **B** 0.07 mm
C 0.007 mm **D** 0.0007 mm

8 Budding in yeast would occur at the most rapid rate in well-fed yeast cells given conditions of

	Temperature (°C)	pH
A	15	7
B	25	5
C	35	7
D	45	5

Items 9, 10 and 11 refer to the accompanying diagram which shows an experiment set up to investigate whether yeast is able to respire anaerobically.

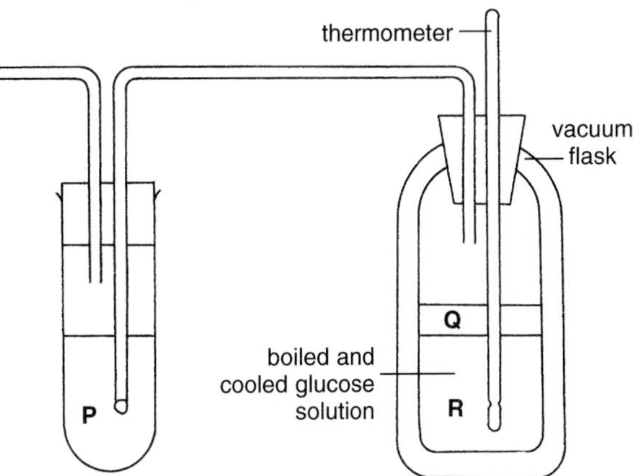

9 The correct positions of the substances involved would be

	P	Q	R
A	oil	indicator	yeast
B	indicator	yeast	oil
C	yeast	oil	indicator
D	indicator	oil	yeast

10 The glucose solution has been boiled in order to

A set up a valid control experiment.
B prevent oxygen being available to the yeast.
C warm the system to an optimum temperature.
D promote production of carbon dioxide.

11 The result obtained when live yeast is used is

	Alcohol produced	Carbon dioxide produced	Heat produced
A	yes	yes	yes
B	no	yes	yes
C	yes	yes	no
D	yes	no	no

12 Before yeast is added to fruit juice during wine-making, rival micro-organisms are destroyed by

A adding sulphur dioxide to the crushed grapes.
B adding carbon dioxide to the crushed grapes.
C boiling the crushed grape mixture.
D filtering the crushed grape mixture.

Questions 13, 14 and 15 refer to the accompanying graph which indicates the results from closely monitoring the changes that take place during the fermentation of a closed batch of wine.

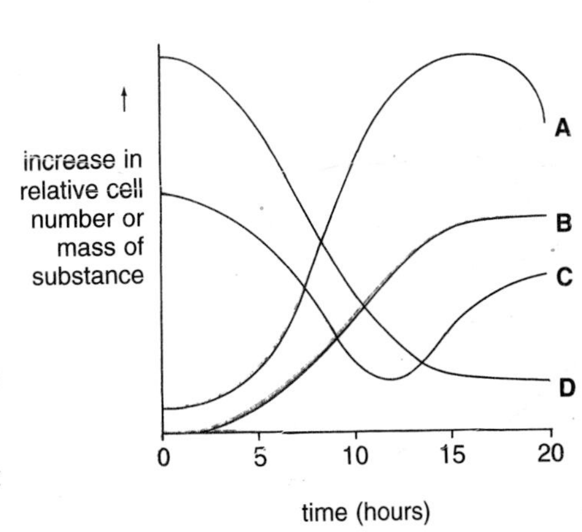

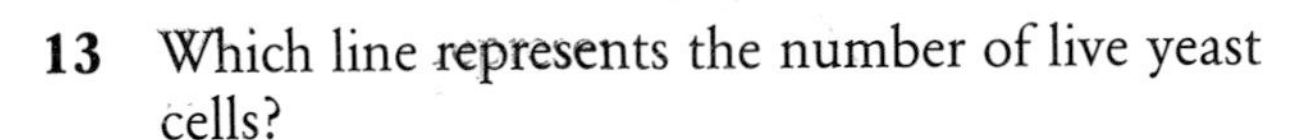

13 Which line represents the number of live yeast cells?

14 Which line represents the mass of glucose present in the fermenter?

15 Which line represents the mass of alcohol present in the fermenter?

16 The accompanying diagram shows one route by which ethanol is produced for use as an alternative fuel.

Which arrow represents fermentation by yeast cells?

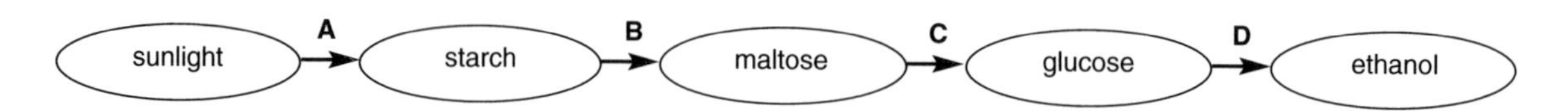

Items 17, 18 and 19 refer to the following information. A multidisc bearing antibiotics 1 to 6 on its side arms was placed on a colony of bacterial species W growing on nutrient agar in a Petri dish. The procedure was repeated for bacterial species X, Y and Z and the four Petri dishes were incubated for two days at 30 °C. The accompanying diagram shows the results.

bacterial species W
bacterial species X
bacterial species Y
bacterial species Z

17 How many bacterial species were resistant to antibiotic 2?

A 0 **B** 1 **C** 2 **D** 3

18 Which bacterial species were sensitive to antibiotic 3?

A W only
B X, Y and Z only
C W, Y and Z only
D W, X, Y and Z

19 Which antibiotic was LEAST effective against bacterial species Y?

A 2 **B** 3 **C** 4 **D** 5

20 Scientists wished to find out if a newly isolated fungus (F) produced an antibiotic. They used a sterile inoculating loop to apply a sample along line F on the nutrient agar in a Petri dish as shown in the accompanying diagram. The procedure was repeated to apply four different species of bacteria (P, Q, R and S) along lines P, Q, R and S.

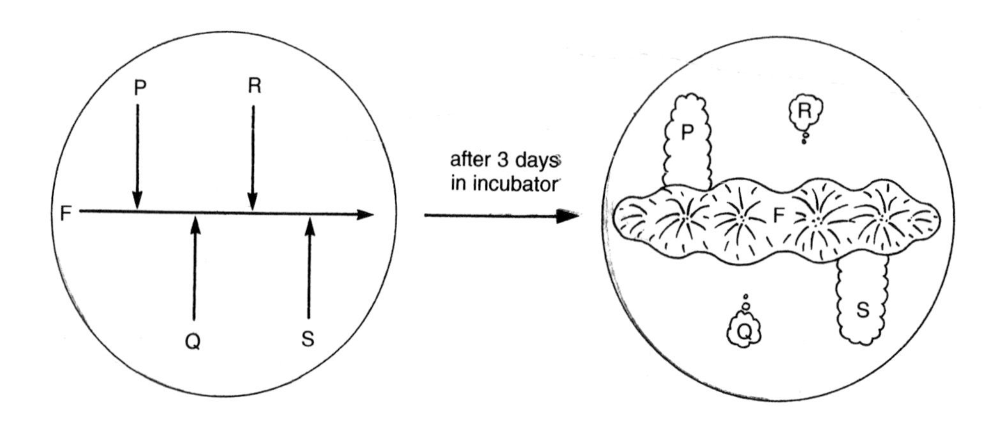

It can be correctly concluded from the results that fungus F

A is resistant to bacteria Q and R and sensitive to P and S.
B is sensitive to bacteria Q and R and resistant to P and S.
C makes an antibiotic to which bacteria P and S are resistant and Q and R are sensitive.
D makes an antibiotic to which bacteria P and S are sensitive and Q and R are resistant.

Items 21, 22 and 23 refer to the following table which shows the effectiveness of four different antibiotics.

Antibiotic	Disease-causing bacterial species			
	1	2	3	4
J	++	–	–	+
K	++	++	–	–
L	++	+	–	++
M	–	–	++	–

++ = very effective   + = effective   – = ineffective

21 Which species of disease-causing bacterium is sensitive to three different antibiotics?

A 1   **B** 2   **C** 3   **D** 4

22 Which two species of disease-causing bacteria are each resistant to two different antibiotics?

A 1 and 3   **B** 2 and 3   **C** 2 and 4   **D** 3 and 4

23 Which antibiotic is effective over the widest range of bacterial species?

A J   **B** K   **C** L   **D** M

24 During the production of yoghurt from milk, yoghurt bacteria

A convert lactose to lactic acid.
B respire aerobically using lactose.
C make milk sugar molecules coagulate.
D raise the pH of milk to an optimum level.

25 Rennin is an enzyme used to make molecules of milk protein coagulate during the production of cheese. The accompanying table shows the results of an investigation into the effect of temperature on the action of rennin.

Temperature (°C)	Time taken for milk to coagulate (min)
10	50
20	40
30	20
40	5
50	60

Which of the following graphs represents these results most accurately?

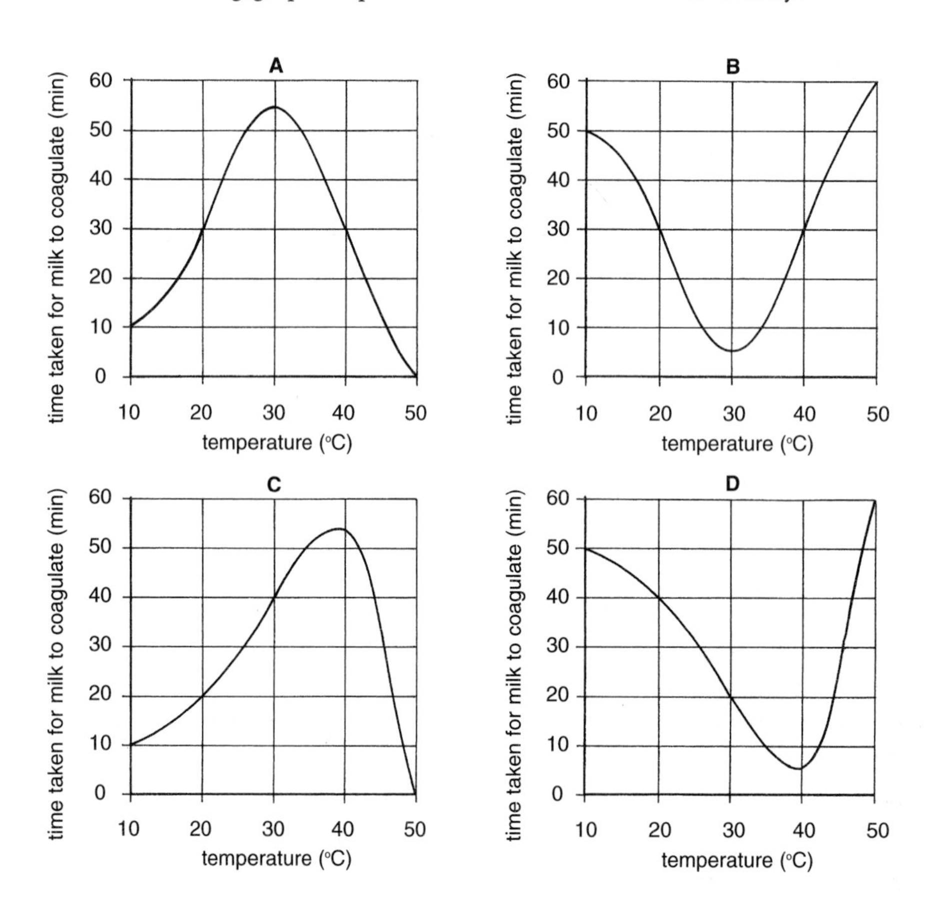

Test 1

1 Diffusion is the movement of molecules of a substance from a region of

A high concentration to a region of low concentration of the same substance.
B low concentration to a region of high concentration of the same substance.
C high concentration to a region of low concentration of a different substance.
D low concentration to a region of high concentration of a different substance.

2 Which of the following diagrams correctly depicts diffusion of oxygen and CO_2 during gas exchange between an air sac and a blood capillary in the mammalian lung?

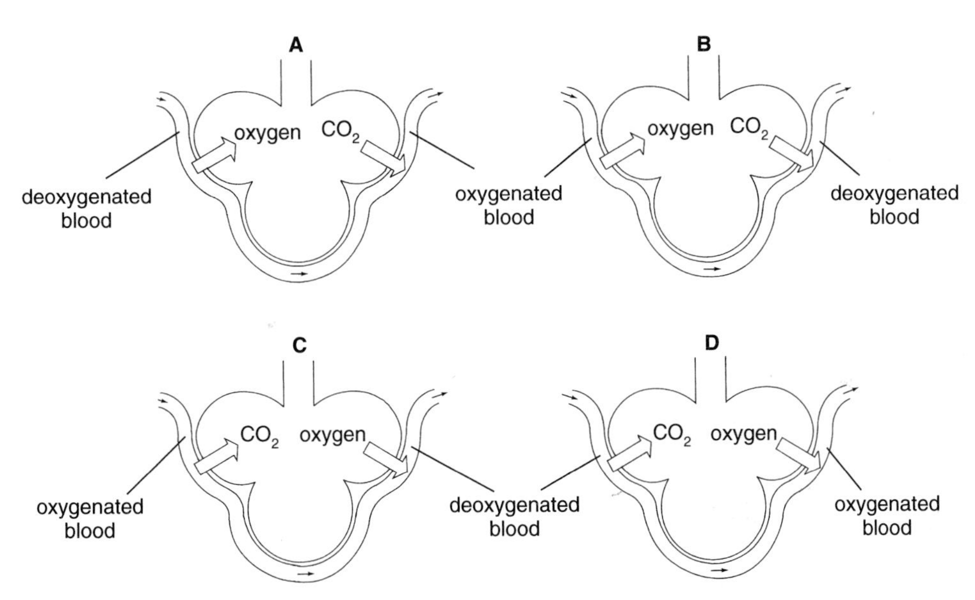

3 The accompanying diagram shows ways in which molecules may move into and out of a respiring animal cell.

Which of these could be diffusion of carbon dioxide molecules?

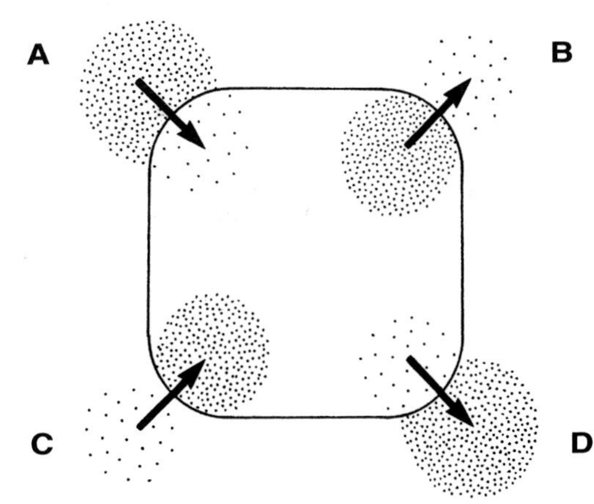

4 Which of the following terms is used to describe a solution with a water concentration higher than that of a comparable solution?

A isotonic **B** hypotonic
C hypertonic **D** plasmolytic

5 Osmosis is the passage through a selectively permeable membrane of

A water from a region of higher solute concentration to a region of lower solute concentration.
B solute from a region of lower water concentration to a region of higher water concentration.
C water from a region of lower solute concentration to a region of higher solute concentration.
D solute from a region of higher water concentration to a region of lower water concentration.

Items 6, 7, 8 and 9 refer to the following information. In an experiment, groups of 100 potato discs were blotted dry and weighed. Each group was then immersed in one of a series of sucrose solutions. After 4 hours, each group was blotted and reweighed. The results are shown in the following table.

Molar concentration of sucrose solution (M)	Initial mass of 100 potato discs (g)	Final mass of 100 potato discs (g)	Gain (+) or loss (−) in mass of 100 potato discs (g)	% gain (+) or loss (−) in mass of 100 potato discs
0.1	20.00	21.00	+1.00	+5.00
0.2	20.00	20.40	+0.40	+2.00
0.3	20.00	19.80	-0.20	-1.00
0.4	22.00	21.12	-0.88	-4.00
0.5	21.00	19.53	-1.47	**box X**

6 Which of the following graphs represents the results for the first four solutions?

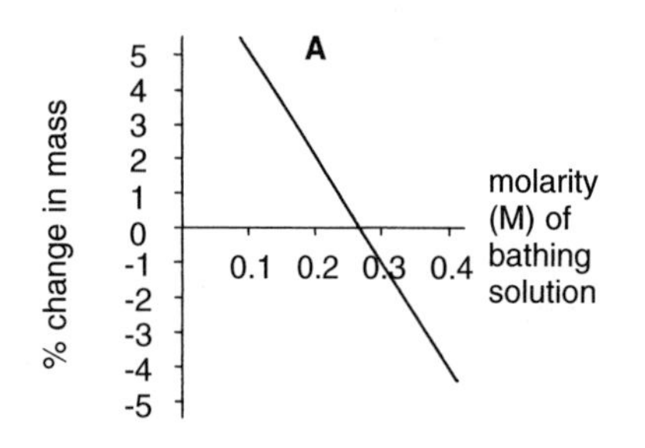

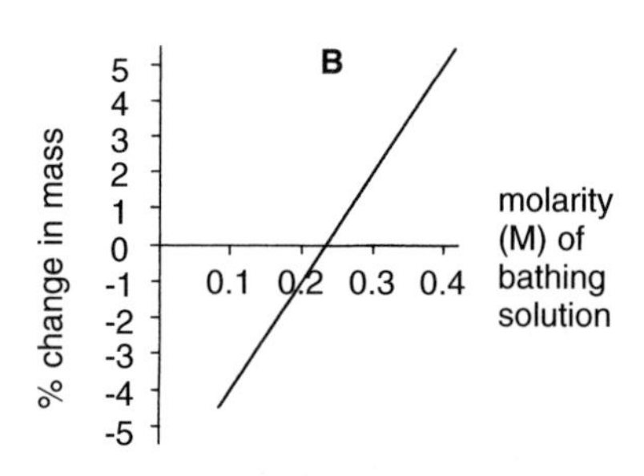

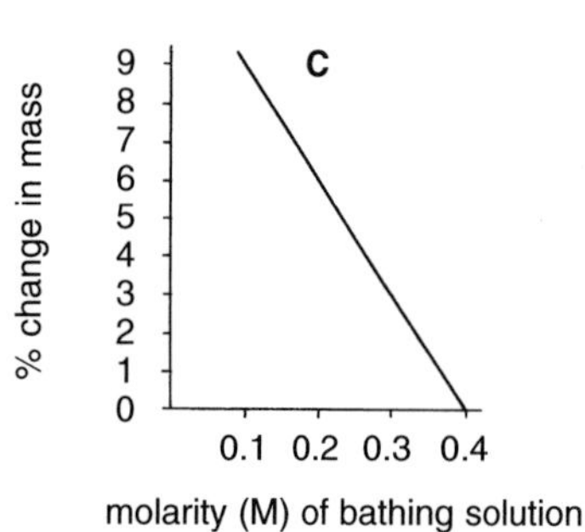

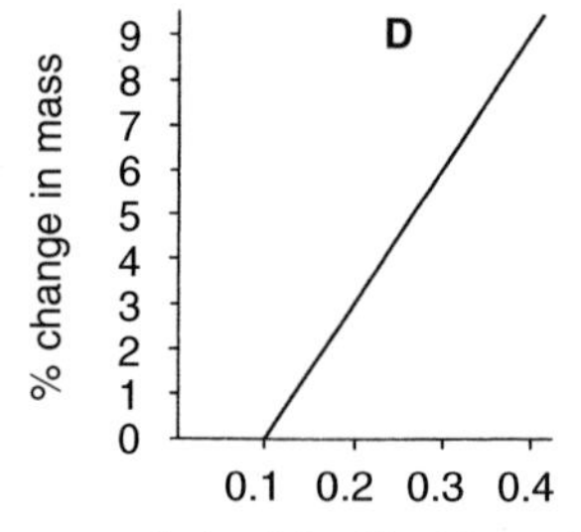

7 Which molarity of sucrose solution has a water concentration closest to that of potato cell sap?

A 0.1M **B** 0.2M **C** 0.3M **D** 0.4M

8 Box X in the table should read

A 7.00. **B** 7.50. **C** −7.00. **D** −7.50.

9 In this experiment it is necessary to convert the results to **percentage** gain or loss in mass in order to

A compensate for the fact that 100 discs were used.
B eliminate the need to repeat the experiment.
C avoid the need to pool results.
D standardise the results.

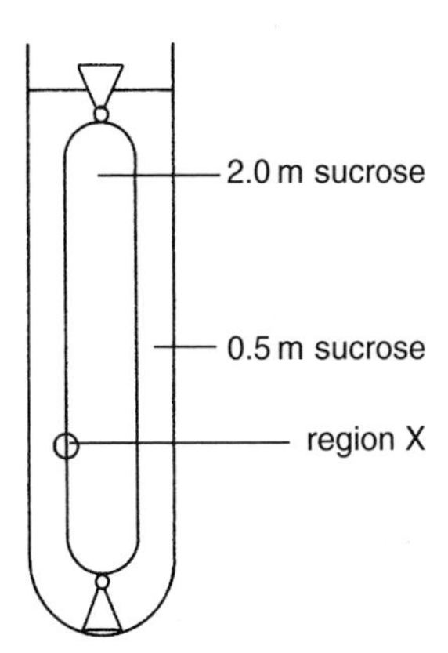

10 The accompanying diagram shows a visking tubing cell model set up to investigate the process of osmosis.

Which of the diagrams below most accurately represents region X at a molecular level?

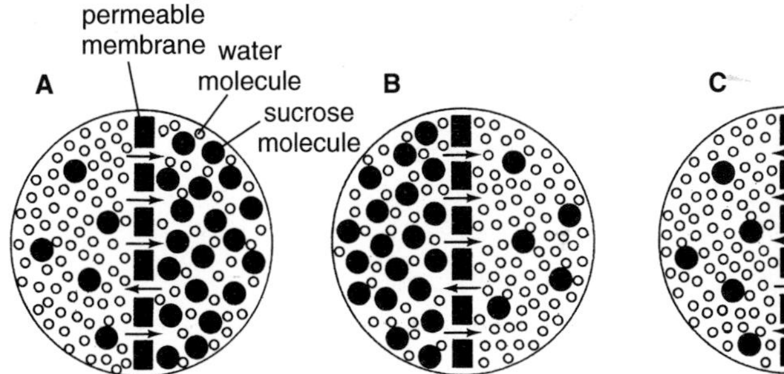

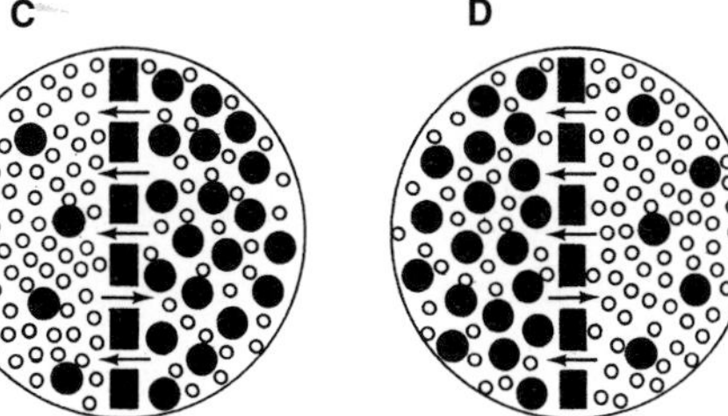

→ = direction of movement of water molecules

11 In a green plant, osmosis is NOT involved in the movement of water molecules from

A the soil solution into a root hair.
B one green leaf cell to another green leaf cell.
C a xylem vessel to a stem cell.
D one xylem vessel to another xylem vessel.

12 The tissue fluid that bathes living cells in the human body is isotonic to 1% salt solution. If human pancreatic cells were immersed in 5% salt solution, they would

A take in water until they were turgid.
B lose water and shrink in size.
C gain water and quickly burst.
D lose salt molecules by diffusion.

13 *Paramecium* has two contractile vacuoles for the evacuation of excess water from its body. The accompanying table records the average time taken by a contractile vacuole to fill up and empty when the animal is immersed in three different liquids X, Y and Z.

Bathing liquid	Time (s)
X	189
Y	28
Z	62

The correct identity of the three bathing liquids is

	Bathing liquid		
	X	Y	Z
A	0.3% salt solution	0.1% salt solution	water
B	0.1% salt solution	water	0.3% salt solution
C	water	0.3% salt solution	0.1% salt solution
D	0.3% salt solution	water	0.1% salt solution

Items 14 and 15 refer to the accompanying diagram showing cells which have been immersed in solutions of different water concentration.

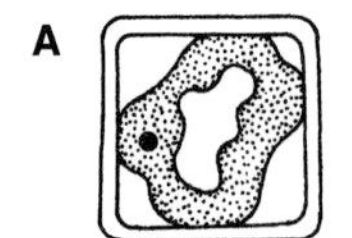

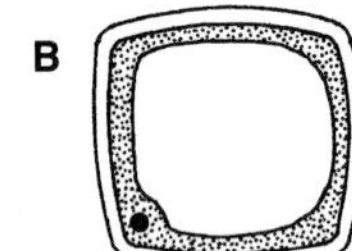

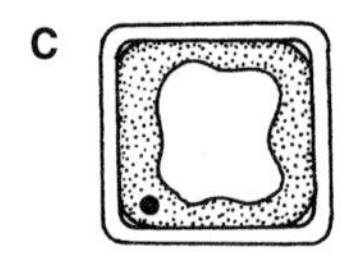

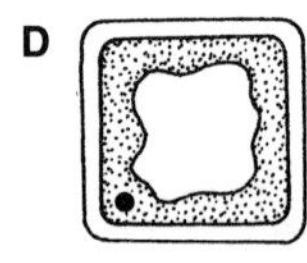

14 In which cell has plasmolysis just begun?

15 Which cell is immersed in the most concentrated sugar solution?

16 If the cells in the accompanying diagram remained in contact as shown, then water would pass by osmosis from BOTH

A R to Q and Q to P.
B Q to S and R to Q.
C P to Q and R to S.
D Q to P and Q to R.

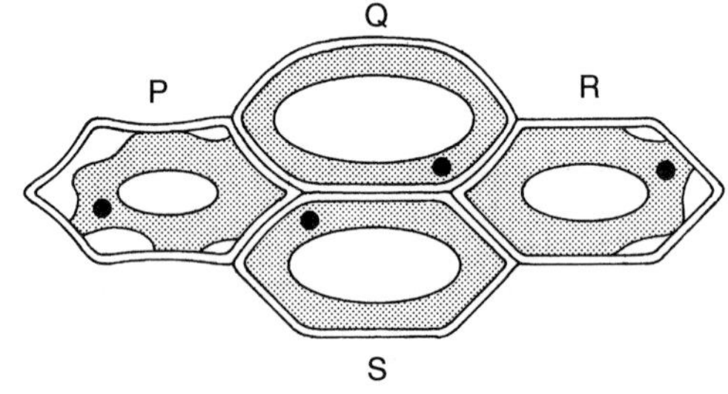

Items 17 and 18 refer to the accompanying diagram which shows the appearance of a plant cell after being immersed in each of four liquids A, B, C and D.

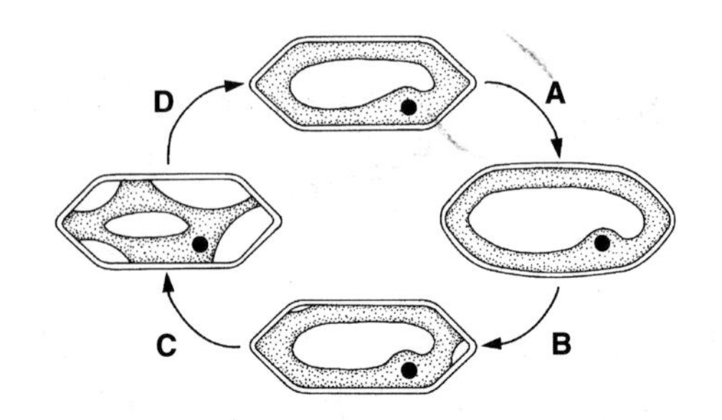

17 Which liquid was most hypotonic to the cell sap?

18 Which liquid was most hypertonic to the cell sap?

Items 19 and 20 refer to the accompanying graph of the results from an experiment where each of four potato cylinders was immersed in a different chemical for a time and then placed in water.

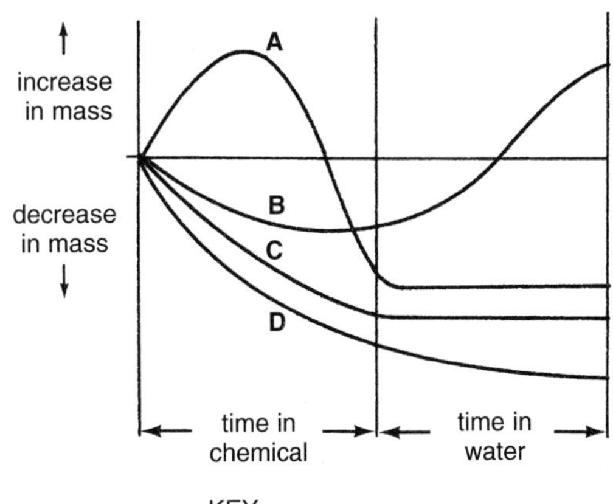

19 Which chemical was NOT toxic (poisonous) to the selectively permeable membranes of the potato cells?

20 Deplasmolysis is the opposite process from plasmolysis. In which cylinder did deplasmolysis take place?

Test 2

1 During diffusion, molecules always move

A across a concentration gradient from low to high concentration.
B down a concentration gradient from high to low concentration.
C up a concentration gradient from low to high concentration.
D against a concentration gradient from high to low concentration.

2 Diffusion is important to the unicellular animal *Amoeba* because it is the means by which

A oxygen, a useful substance, enters and CO_2, a waste product, leaves.
B CO_2, a useful substance, enters and oxygen, a waste product, leaves.
C oxygen, a waste product, enters and CO_2, a useful substance, leaves.
D CO_2, a waste product, enters and oxygen, a useful substance, leaves.

Items 3, 4, 5 and 6 refer to the accompanying diagram which shows five liquids (P, Q, R, S and T) and molecular close-ups of them.

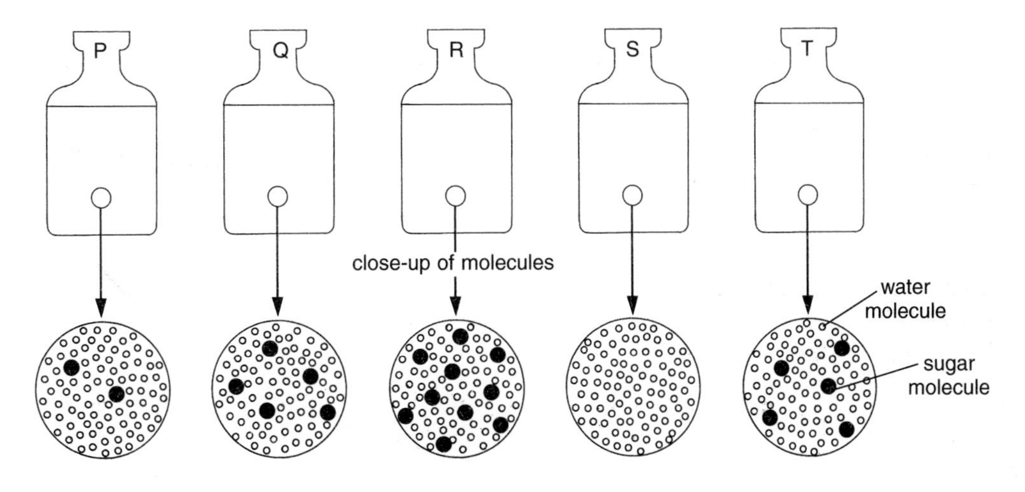

3 The liquid which is hypertonic to all of the others is

A P. **B** Q. **C** R. **D** S.

4 The liquid which is hypotonic to all of the others is

A P. **B** Q. **C** R. **D** S.

5 Liquid T is isotonic to

A P. **B** Q. **C** R. **D** S.

6 It is correct to say that liquids

A P and Q are hypotonic to both S and T.
B P and Q are hypertonic to both S and T.
C R and T are hypotonic to both P and S.
D R and T are hypertonic to both P and S.

7 Osmosis can be defined as the net flow of water molecules through a selectively permeable membrane from a region of

A high water concentration to a region of lower water concentration.
B high solute concentration to a region of lower solute concentration.
C low water concentration to a region of higher water concentration.
D low solute concentration to a region of isotonic solute concentration.

Items 8 and 9 refer to the accompanying diagram. It shows the results of an experiment where turnip cylinders, initially measuring 5 cm in length, were immersed in three different liquids for 24 hours. The test tubes were kept in a thermostatically controlled water-bath during this time.

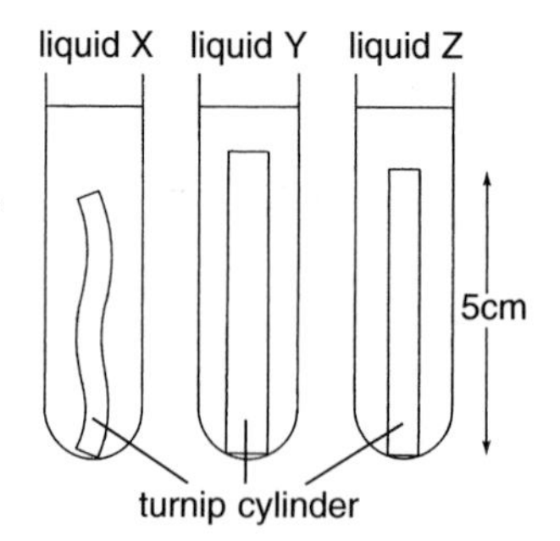

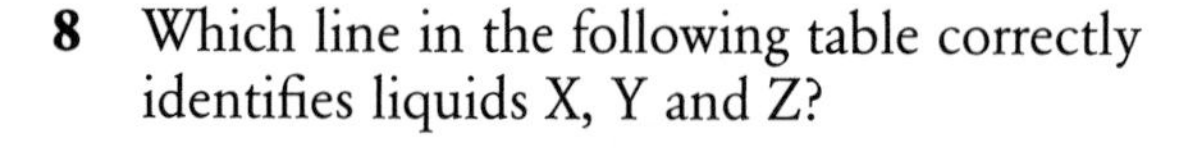

8 Which line in the following table correctly identifies liquids X, Y and Z?

	Bathing liquid		
	X	**Y**	**Z**
A	I molar sucrose	0.3 molar sucrose	pure water
B	I molar sucrose	pure water	0.3 molar sucrose
C	0.3 molar sucrose	I molar sucrose	pure water
D	0.3 molar sucrose	pure water	I molar sucrose

9 The factor that was varied in this experiment was

A temperature of bathing liquid.
B length of turnip cylinder.
C concentration of sucrose solution.
D diameter of turnip cylinder.

Items 10 and 11 refer to the accompanying diagram which shows an experiment set up to investigate diffusion of molecules.

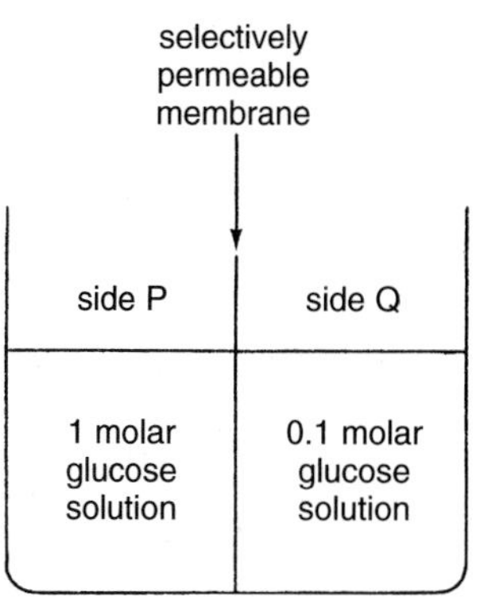

10 Water molecules will move by osmosis through the membrane

A from side P to side Q only.
B from side Q to side P only.
C in both directions but mainly from P to Q.
D in both directions but mainly from Q to P.

11 Over a longer period of time, glucose molecules will diffuse

A from side P to side Q only.
B from side Q to side P only.
C in both directions but mainly from P to Q.
D in both directions but mainly from Q to P.

Items 12 and 13 refer to the experiment shown in the accompanying diagram.

12 After a few days, which of the following will have occurred?

A a rise in level X and a drop in level Y
B a drop in level X and a drop in level Y
C a rise in level X and a rise in level Y
D a drop in level X and a rise in level Y

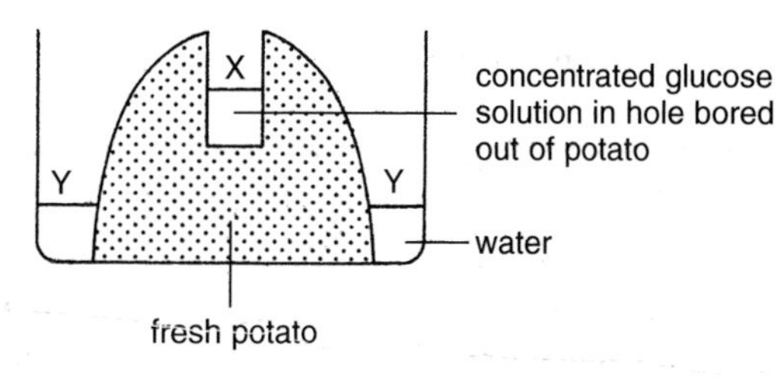

13 Which of the following diagrams shows a suitable control for the above experiment?

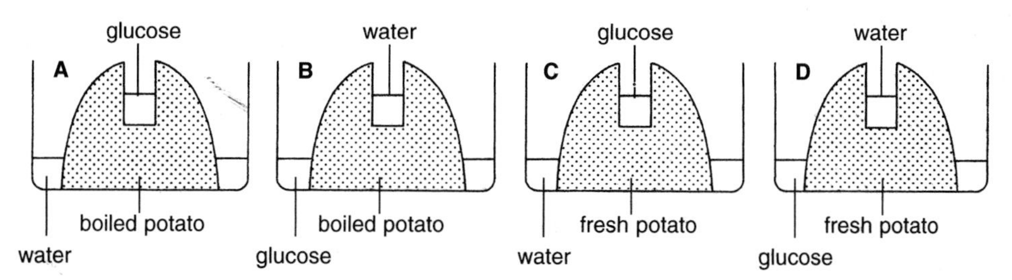

Items 14 and 15 refer to both the accompanying diagram of a normal red blood cell immersed in 1% salt solution and to the possible answers illustrated in the second diagram.

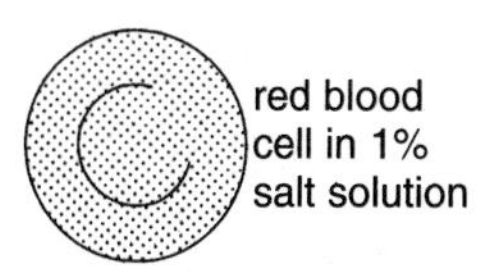

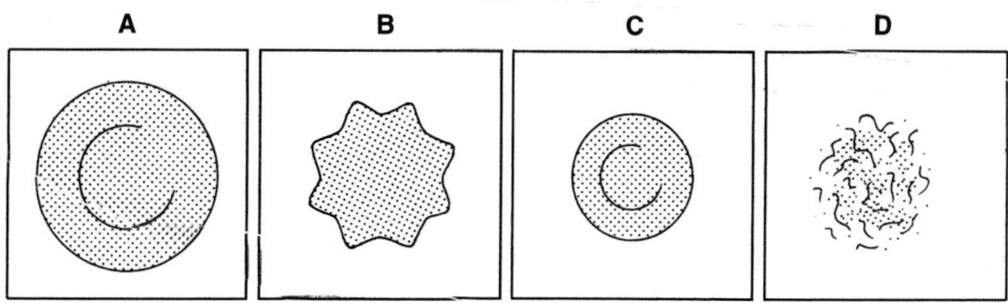

14 Which part of the second diagram shows the appearance of a red blood cell immersed in water?

15 Which part of the second diagram shows the appearance of a red blood cell immersed in 2% salt solution.

16 *Stentor* is a unicellular animal which lives in pond water isotonic to 0.1% salt solution. It uses its two contractile vacuoles to remove excess water from its body.

If *Stentor* were moved from its normal environment to pure water, its contractile vacuoles would

A stop working until normal conditions returned.
B continue to work at their normal rate.
C increase their rate of emptying.
D decrease their rate of emptying.

17 The accompanying diagram shows the appearance of a plant cell immersed in a solution which is isotonic to the cell's sap.

Which of the diagrams shown below most accurately represents the appearance of this cell after immersion in a hypertonic solution?

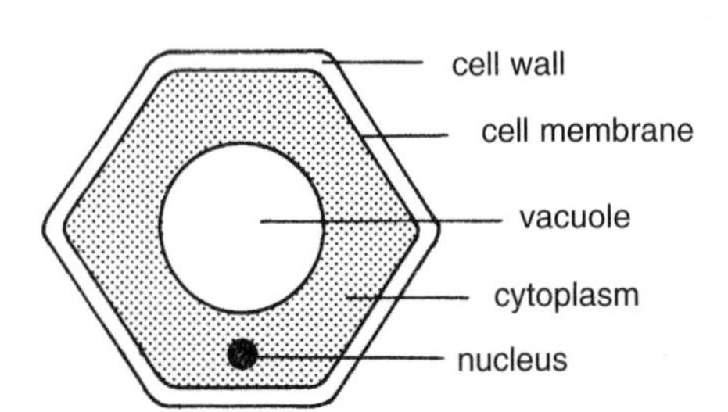

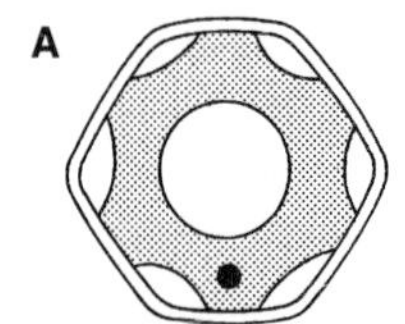

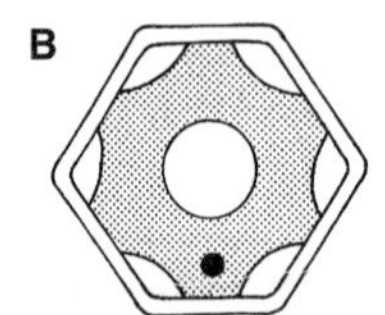

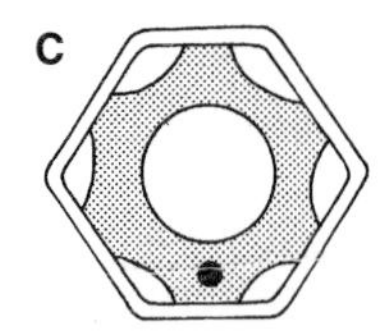

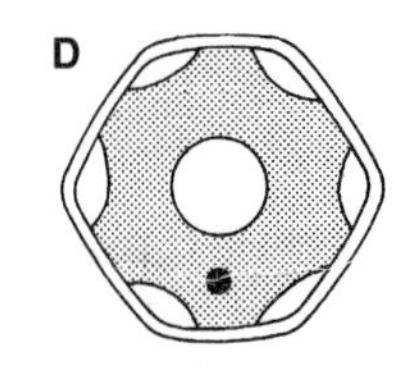

Items 18 and 19 refer to the following possible answers.

A turgid **B** flaccid **C** plasmolysed **D** deplasmolysed

18 Which term is used to describe a plant tissue swollen with water?

19 Which term is used to describe a plant cell whose contents have shrunk and pulled away from the cell wall following excessive water loss?

20 The accompanying diagram shows four red onion cells immersed in four different solutions P, Q, R and S.

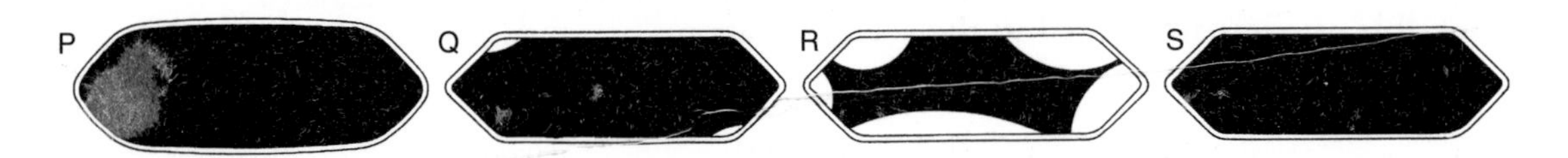

If the bathing solutions were arranged in order of increasing water concentration, the sequence would be

A R, Q, S, P. **B** R, S, Q, P. **C** P, Q, S, R. **D** P, S, Q, R.

3 Enzyme action

Test 1

1 Which of the following graphs refers to a chemical reaction controlled by a catalyst?

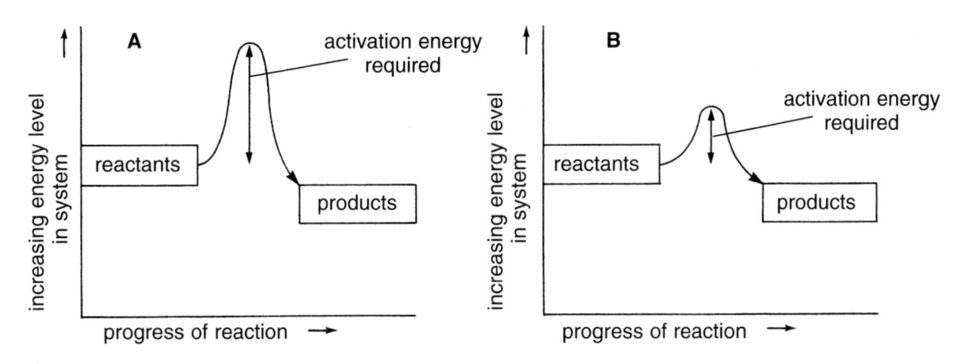

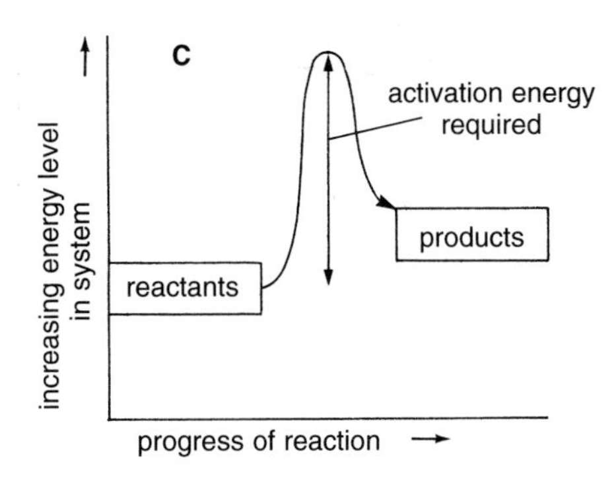

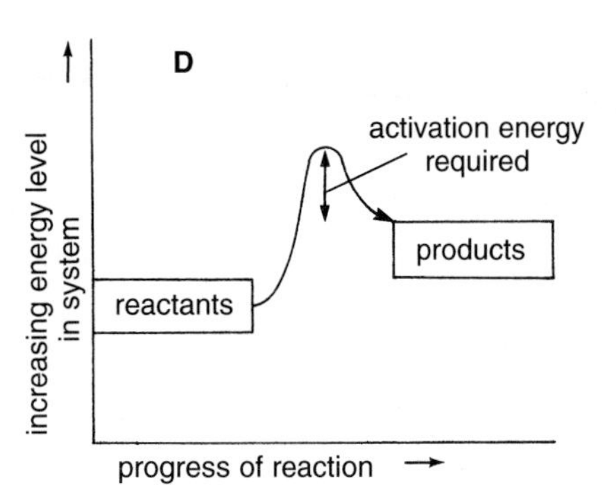

2 Each enzyme is a

A protein molecule produced by living cells to catalyse a particular biochemical reaction.
B living molecule produced by cells to digest food in the alimentary canal.
C protein molecule produced by living cells to catalyse a variety of biochemical reactions.
D living molecule produced by cells to synthesise complex substances from simpler ones.

3 The following three statements refer to enzymes.

1 After promoting a biochemical reaction, an enzyme's molecular structure remains unaltered.

2 The product of an enzyme-controlled reaction is called the enzyme-substrate complex.

3 An enzyme's molecular shape is complementary to that of its substrate.

Which of these statements is CORRECT?

A 1 only **B** 1 and 3 only
C 2 and 3 only **D** 1, 2 and 3

Items 4, 5 and 6 refer to the experiment shown in the accompanying diagram set up to investigate the effect of catalase on the breakdown of hydrogen peroxide.

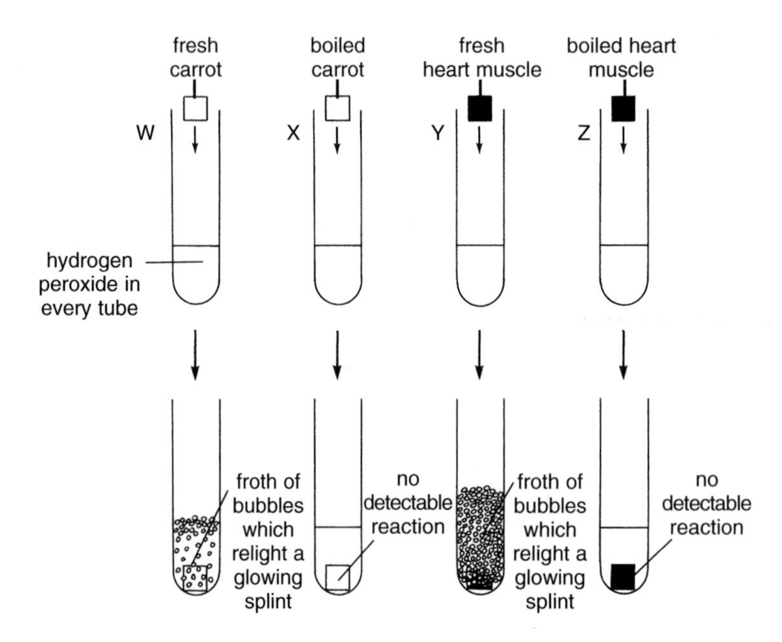

4 The two test tubes that received denatured catalase were

A W and X. **B** W and Y.
C X and Y. **D** X and Z.

Items 5 and 6 also refer to the following possible answers.

A oxygen **B** carbon dioxide
C heart muscle **D** hydrogen peroxide

5 Which substance is an end product of the reaction?

6 Which substance is the substrate?

7 In an enzyme experiment set up to investigate the breakdown of protein, 878 µg of amino acids were produced in 2.25 hours. Expressed in µg/min, the rate of the reaction was

A 6.1. **B** 6.5. **C** 390.2. **D** 1975.5.

Items 8, 9 and 10 refer to the following diagram of enzyme action.

8 Which molecules make up the substrate?

A 1 and 2 **B** 2 and 3
C 4 and 5 **D** 7 and 8

9 Which molecule is the product?

A 2 **B** 3 **C** 7 **D** 8

10 Which of the following depict the enzyme–substrate complex?

A 1, 2 and 3 **B** 2 and 3
C 4, 5 and 6 **D** 7 and 8

11 A molecule of maltose sugar is composed of two molecules of glucose. In the first of the accompanying diagrams, amylase is about to promote the digestion of starch to maltose.

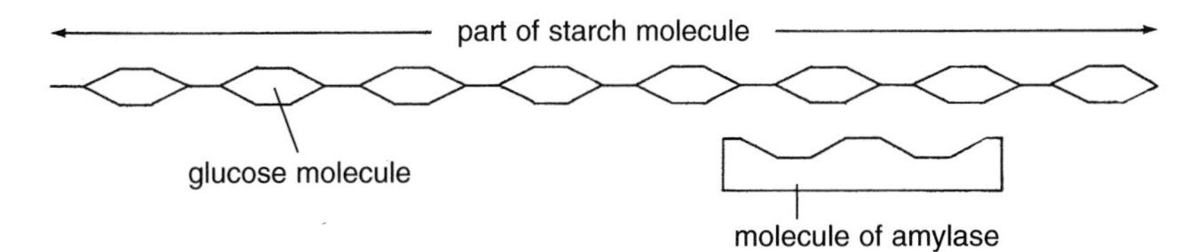

Which of the following diagrams best represents molecules at the end of the reaction?

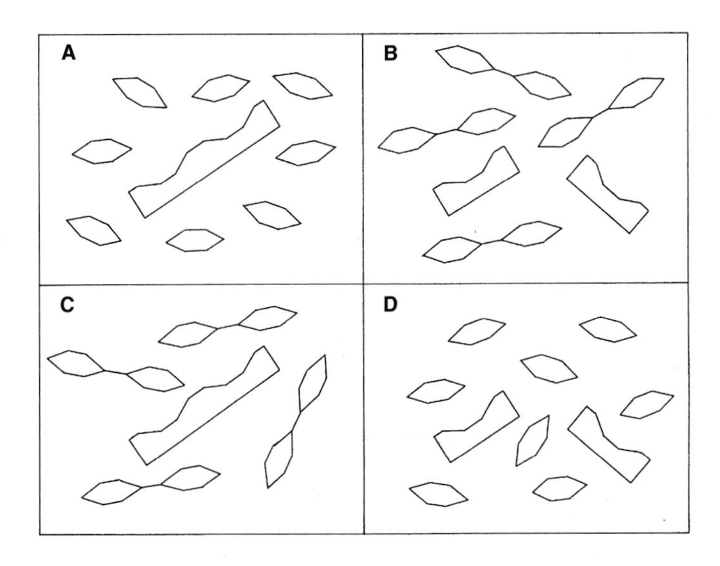

Items 12 and 13 refer to the experiment in the accompanying diagram. The test tubes were kept at 37 °C in a water-bath and their contents were maintained at pH 7 using buffer solution.

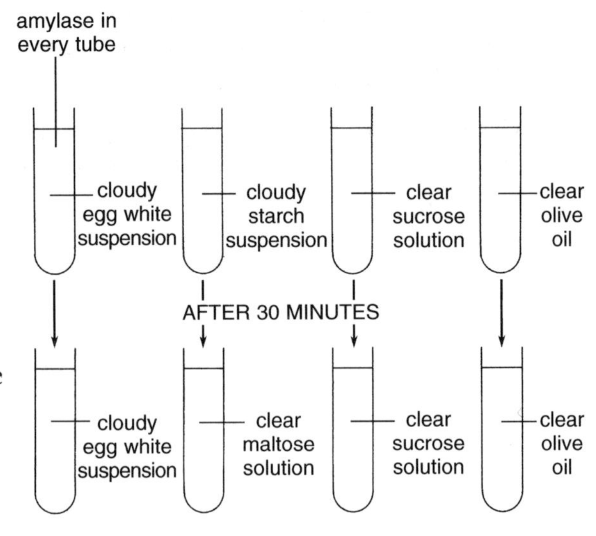

12 The variable factor investigated in this experiment was the

A substrate. **B** enzyme.
C pH. **D** temperature.

13 It can be correctly concluded that amylase

A digests all the substrates except egg white.
B is most active at body temperature.
C works best at neutral pH.
D is specific to its substrate.

14 A group of students wished to investigate the hypothesis that soaked oat grains and soaked barley grains contain amylase. They set up a Petri dish as shown in the accompanying diagram. Each seed was cut in half and the two halves placed cut surface down on the starch agar.

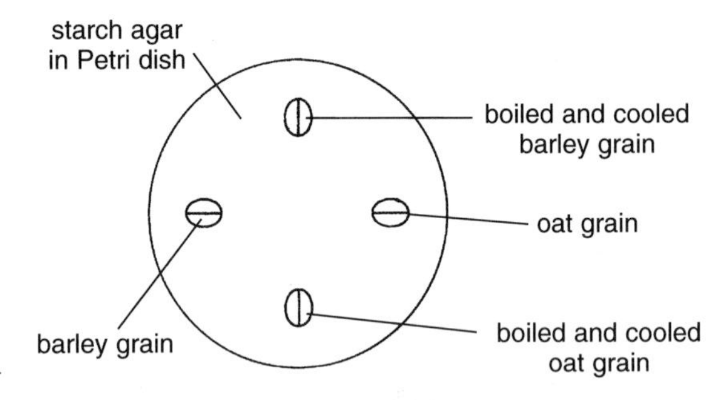

After 24 hours the seeds were removed and the dish flooded with iodine solution.

Which of the results shown in the following diagram supports the students' hypothesis?

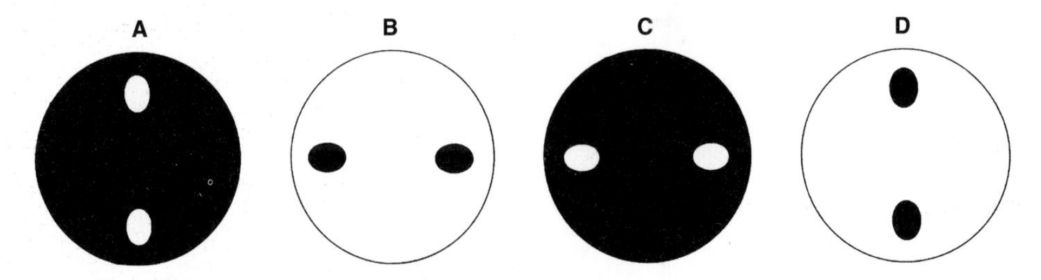

15 A control is set up to

A increase the reliability of the results.
B make the experiment fair in every way.
C ensure that the results are accurate.
D show that the results are due to the factor being investigated.

16 The accompanying diagram shows an experiment set up to investigate the action of amylase on starch.

When the food tests for simple sugar and starch are carried out, a positive result is obtained in BOTH tubes

A 1 and 3. **B** 1 and 4.
C 2 and 3. **D** 3 and 4.

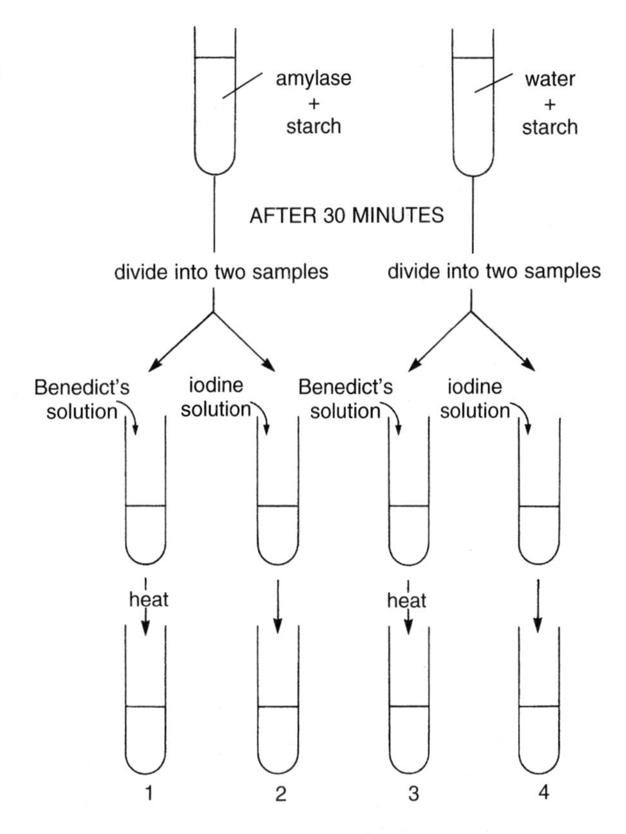

Items 17 and 18 refer to the accompanying graph which shows the rate of an enzyme's activity at different temperatures.

17 The greatest increase in rate of enzyme activity occurred between temperatures

A 20 and 25 °C.
B 25 and 30 °C.
C 30 and 35 °C.
D 35 and 40 °C.

18 The greatest change in enzyme activity took place between temperatures

A 15 and 25 °C.
B 25 and 35 °C.
C 35 and 45 °C.
D 45 and 55 °C.

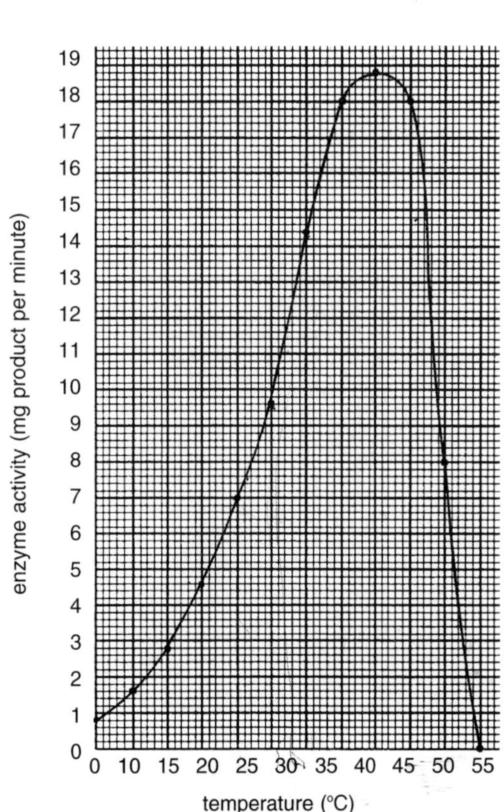

Items 19 and 20 refer to the following information. Albumin is a protein found in egg white. A protease is a protein-digesting enzyme.

To investigate the effect of pH on the activity of two proteases, X and Y, on albumin, the experiment shown in the accompanying diagram was carried out.

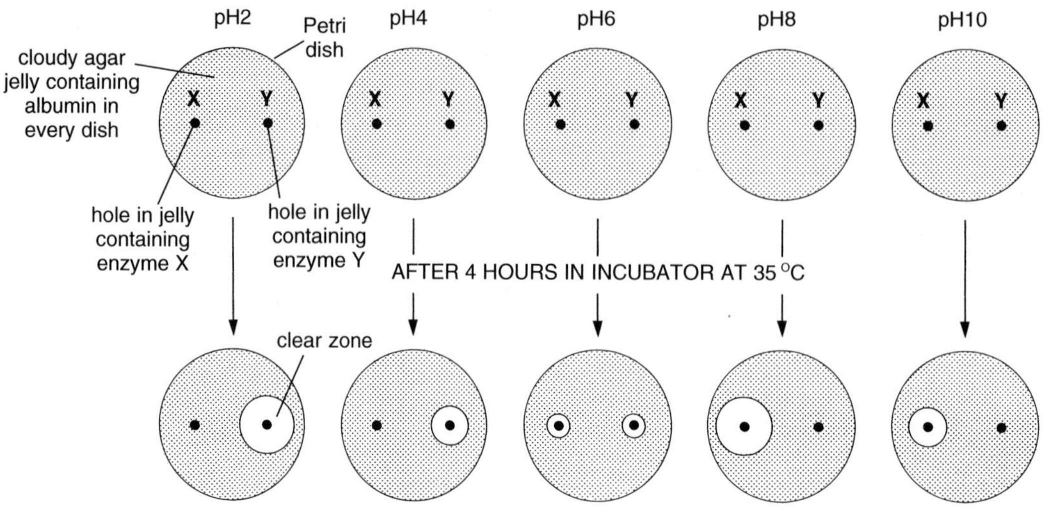

19 The enzymes were equally active at pH

A 2. **B** 4. **C** 6. **D** 8.

20 The optimum pH for enzyme X is

A 2. **B** 4. **C** 8. **D** 10.

21 The graph in the accompanying diagram shows the effect of pH on the activity of three enzymes, X, Y and Z.

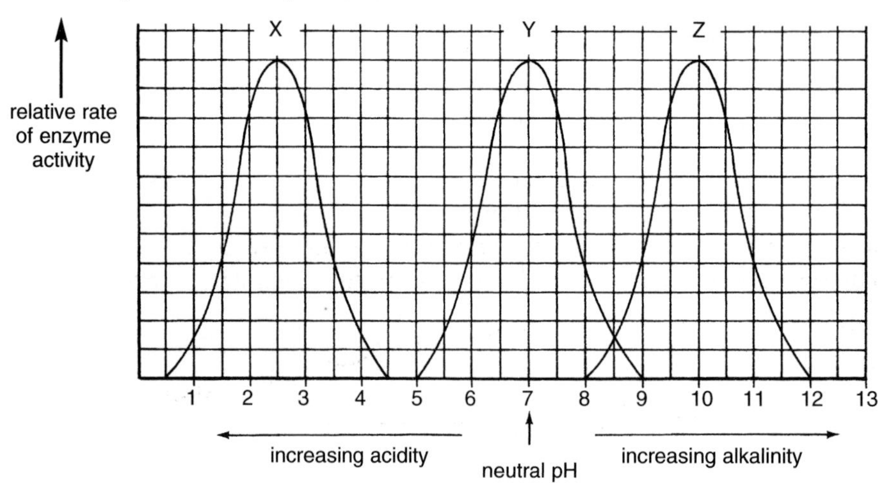

Which of the following conclusions CANNOT be drawn from the graph?

A The region of pH scale at which each enzyme shows maximum activity is similar.
B Two of the enzymes can be found working in alkaline conditions.
C The breadth of working range of pH of each of the enzymes is equal.
D The optimum pH value for each enzyme differs significantly from that of the others.

Items 22, 23, 24 and 25 refer to the following information. An unusual enzyme has been extracted from a species of bacterium. The accompanying table shows the percentage of protein that is digested by this enzyme at different temperatures and levels of pH.

		Percentage digestion at temperature (°C) of:			
		40	**50**	**60**	**70**
Percentage digestion at pH of:	**5**	12	19	31	2
	7	61	73	88	21
	9	67	85	97	33
	11	16	25	42	3

22 The optimum pH for the activity of this enzyme is

A 5. **B** 7. **C** 9. **D** 11.

23 The optimum temperature (in °C) for the activity of this enzyme is

A 40. **B** 50. **C** 60. **D** 70.

24 Which of the following would be the MOST LIKELY percentage of protein digested by this enzyme at pH 8 and temperature 55 °C?

A 68 **B** 71 **C** 86 **D** 99

25 By how many times is the percentage of protein digested at 60 °C and pH 11 greater than that digested at 40 °C and pH 5?

A 3.5 **B** 30.0 **C** 54.0 **D** 505.0

Test 2

1 It is INCORRECT to say that a catalyst

A speeds up the rate of a chemical reaction.
B takes part in the chemical reaction that it promotes.
C lowers the energy input needed to make the chemical reaction proceed.
D undergoes a change in molecular structure as a result of the chemical reaction.

2 The accompanying diagram shows a molecule of an enzyme.

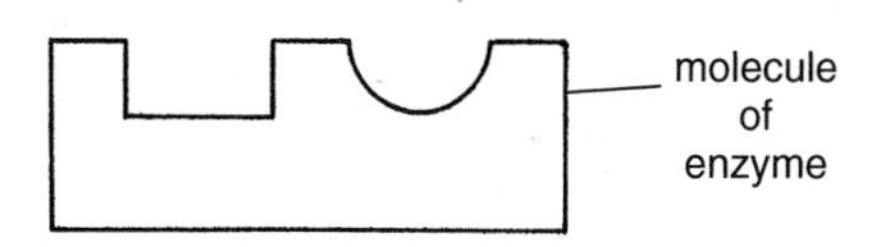

Which of the following diagrams shows its substrate?

A

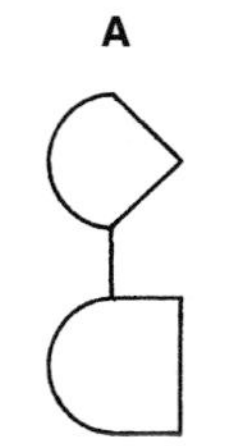

B

C

D

3 An enzyme molecule is

A responsible for catalysing several types of biochemical reaction.
B unaffected by changes in temperature.
C composed of complex carbohydrate.
D able to perform its role repeatedly.

Questions 4, 5 and 6 refer to the following possible answers.

A optimum **B** specific
C denatured **D** synthetic

4 Which term is used to describe an enzyme that has been destroyed?

5 Which term is used to describe the condition of a factor at which an enzyme is most active?

6 Which term is used to describe the lock-and-key fit that takes place between a molecule of an enzyme and its substrate?

Items 7, 8 and 9 refer to the experiment shown in the accompanying diagram and to the table of results that follows it.

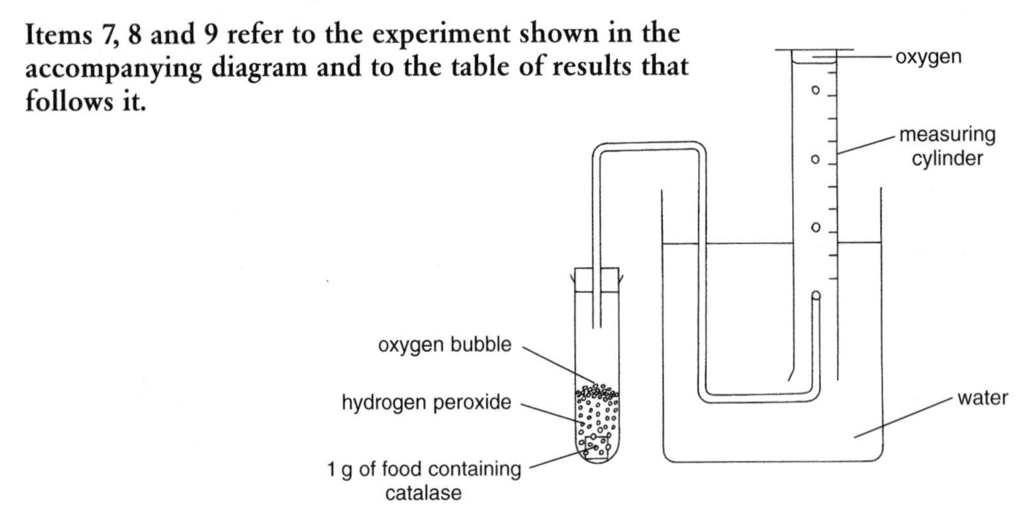

Trial number	Time required to collect 1 cm³ of oxygen (s)			
	carrot	potato	liver	kidney
1	133	87	40	58
2	147	91	42	61
3	129	89	39	57
4	151	107	**see item 8**	60
5	165	96	45	64
average	145	**see item 7**	42	60

7 The average time (in seconds) required to collect 1 cm^3 of oxygen when using potato as the source of catalase was

A 47. **B** 74. **C** 94. **D** 470.

8 The time (in seconds) required to collect 1 cm^3 of oxygen during trial number 4 using liver was

A 42. **B** 44. **C** 166. **D** 210.

9 It can be correctly concluded from this experiment that per unit mass

A potato contains less catalase than kidney.
B carrot contains more catalase than potato.
C potato and carrot both contain more catalase than liver.
D carrot and liver both contain less catalase than kidney.

10 The accompanying diagram shows three stages that occur during an enzyme-controlled reaction.

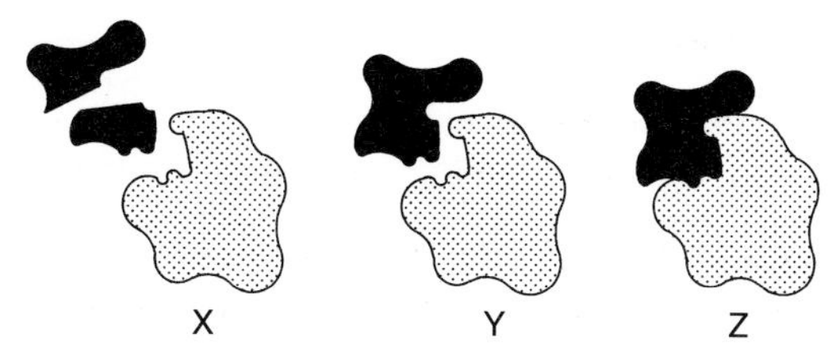

Which of the following indicates the correct sequence in which the three stages would occur if the enzyme promotes the building-up of a complex molecule from simpler ones?

A X → Y → Z **B** Y → Z → X
C X → Z → Y **D** Y → X → Z

11 The enzyme referred to in question 10 is shown again in the accompanying diagram. Which arrow represents the enzyme's active site?

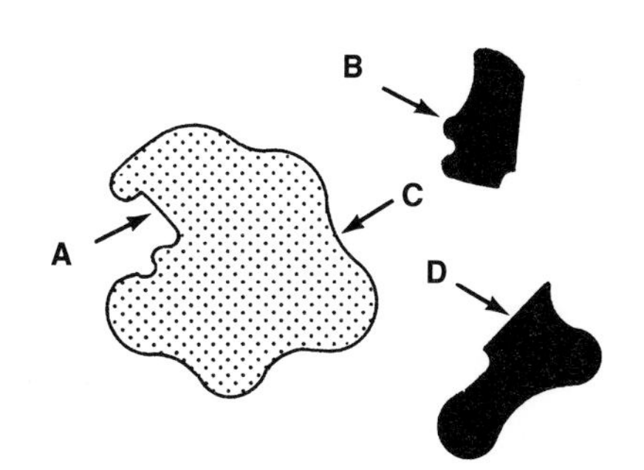

12 The accompanying diagram shows an experiment set up to investigate the effect of the enzymes in a snail's gut on starch molecules.

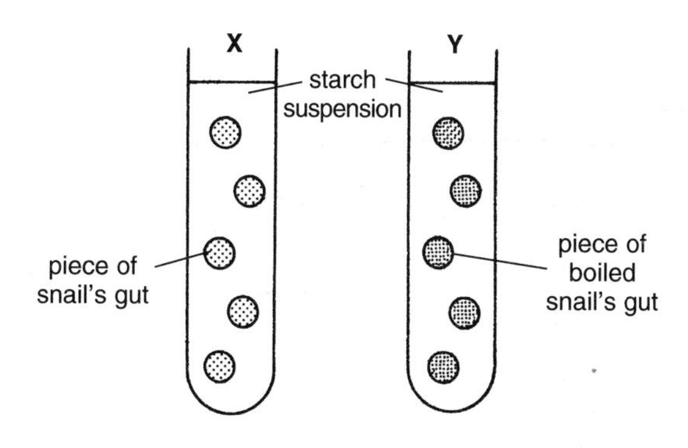

Which of the following sets of results would be obtained if the snail's gut contained amylase?

A

	At start		After 2 hours	
	Starch test	Simple sugar test	Starch test	Simple sugar test
X	+	–	+	–
Y	+	–	–	+

B

	At start		After 2 hours	
	Starch test	Simple sugar test	Starch test	Simple sugar test
X	+	–	–	+
Y	–	+	–	+

C

	At start		After 2 hours	
	Starch test	Simple sugar test	Starch test	Simple sugar test
X	–	+	+	–
Y	–	+	–	+

D

	At start		After 2 hours	
	Starch test	Simple sugar test	Starch test	Simple sugar test
X	+	–	–	+
Y	+	–	+	–

Items 13, 14 and 15 refer to the accompanying diagram which shows an experiment set up to investigate the action of potato phosphorylase on glucose-1-phosphate (G-1-P).

cavity tile

G-I-P + phoshorylase in each cavity

row Y

G-I-P + distilled water in each cavity

13 When the tile is being set up, the cavities in row Y should receive

A distilled water + phosphorylase.
B phosphorylase + glucose-1-phosphate.
C distilled water + starch.
D starch + phosphorylase.

14 The results of this experiment are obtained by adding iodine solution to the mixtures in the cavities at 3-minute intervals.

In the accompanying diagram each number represents the time (in minutes from the start) when iodine should be added to the cavity. Which tile shows the CORRECT procedure?

A

0 0 0 0
3 3 3 3
6 6 6 6

B

0 3 6 9
0 3 6 9
0 3 6 9

C

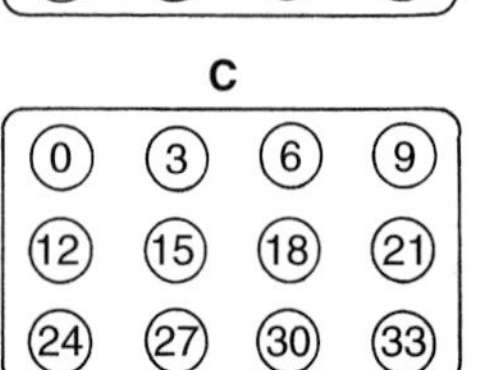

D

0 3 6 9
3 6 9 12
6 9 12 15

15 Which tile in the accompanying diagram shows the results that are obtained once the correct procedure has been followed?

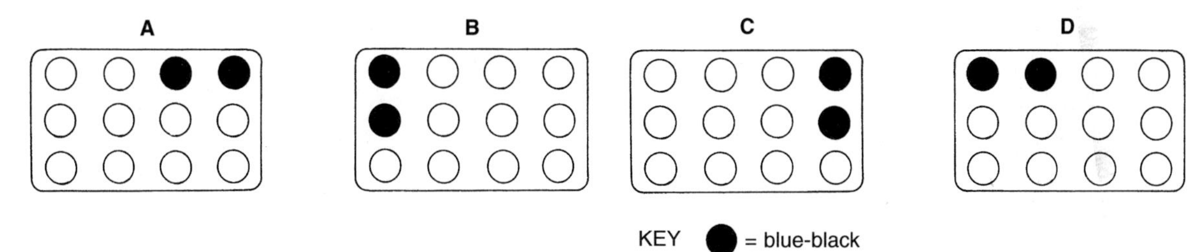

16 The accompanying diagram shows an investigation into the effect of temperature on the activity of lipase.

Temperature of tube's contents during experiment	15 °C	35 °C	55 °C
Colour of tube's contents at start	green	green	green

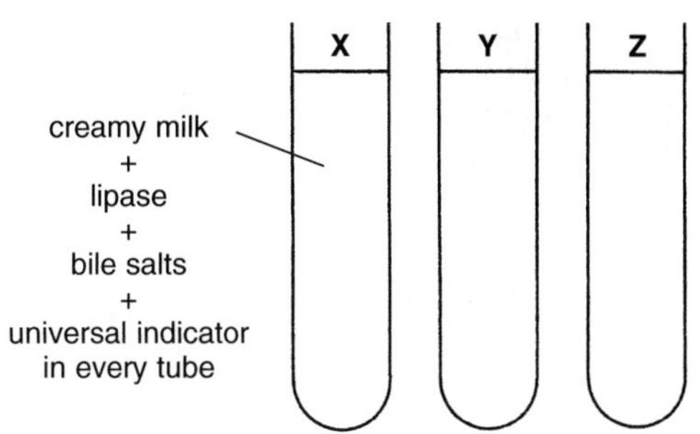

After 30 minutes, the colours of the contents of the tubes will be

	X	Y	Z
A	green	orange	yellow
B	yellow	green	orange
C	orange	yellow	green
D	yellow	orange	green

17 Which of the following graphs shows the effect of temperature on the activity of an enzyme from the human body?

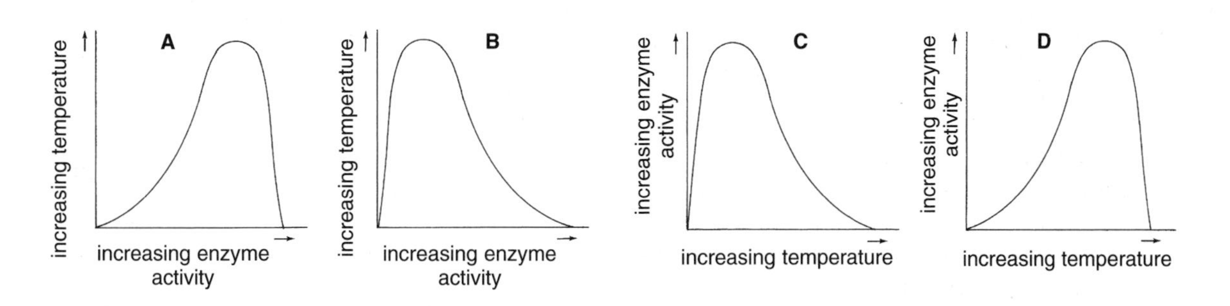

18 Diastase (plant amylase) is an enzyme which digests starch to maltose.

The accompanying diagram shows a Petri dish of starch agar set up to investigate the effect of pH on the activity of diastase.

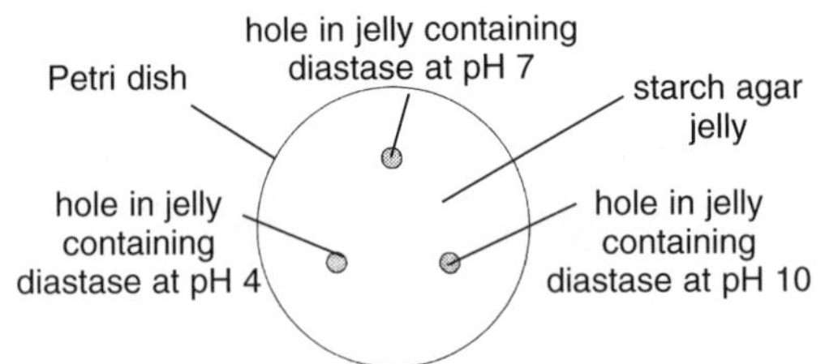

After 24 hours at room temperature, the plate was flooded with iodine solution. Which of the following diagrams indicates that diastase is most active at pH 7?

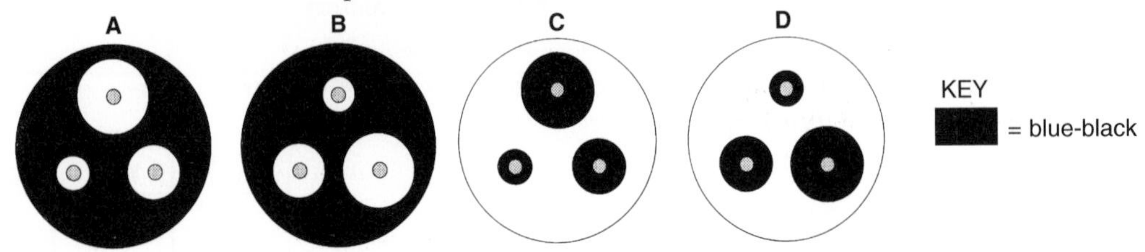

Items 19, 20 and 21 refer to the following information. Before being used in an experiment, bovine catalase solution, hydrogen peroxide solution and a set of pH buffers were incubated at 25 °C. Uniform volumes of these liquids were then used to investigate the effect of pH on the activity of catalase. The results are shown in the following table.

	pH						
	6	7	8	9	10	11	12
Time taken to collect 1 cm^3 of oxygen (s)	108	98	57	48	60	86	96

19 Which of the following graphs CORRECTLY represents the data in the table?

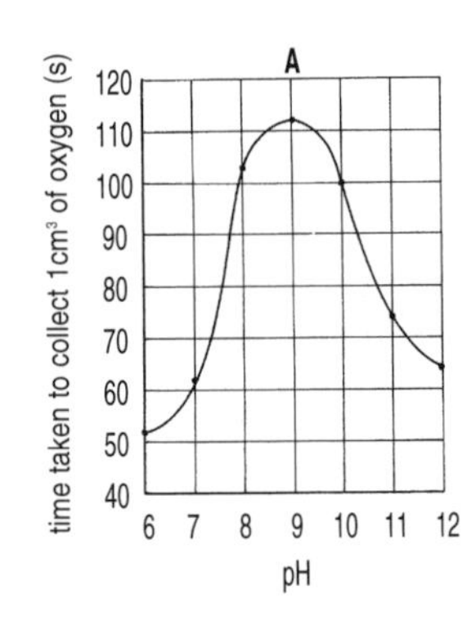

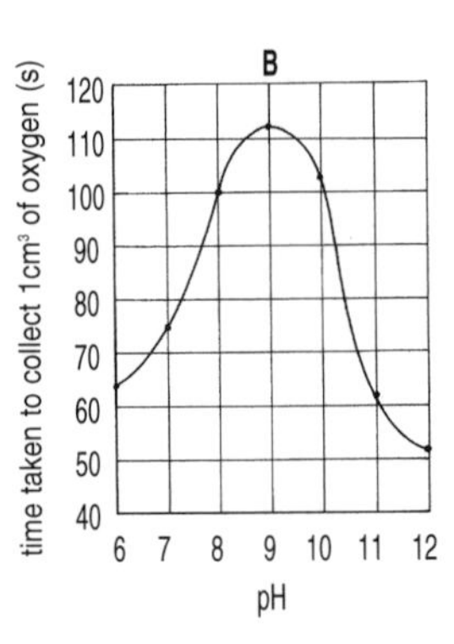

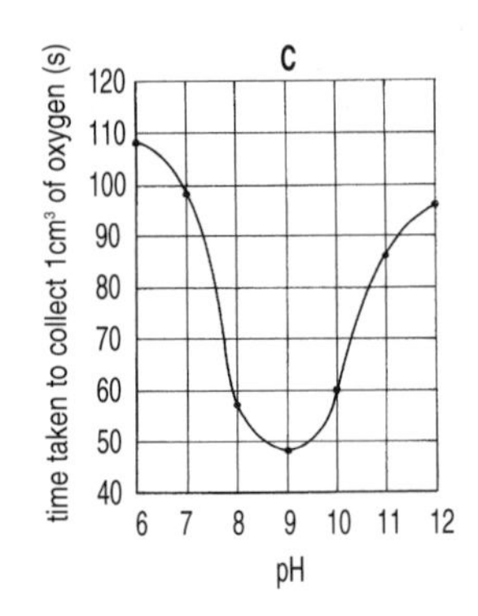

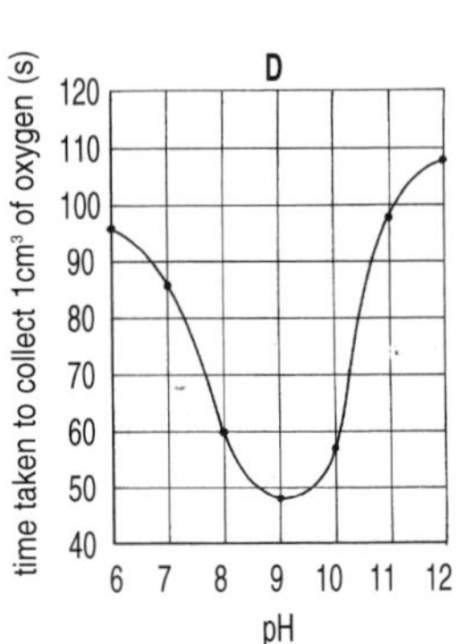

20 The enzyme was most active over the pH range

A 6–8. **B** 7–9. **C** 8–10. **D** 9–11.

21 Pre-incubation of the chemicals used in this experiment is necessary to ensure that

A the experiment contains a valid set of controls.
B temperature is the only variable factor under investigation.
C the results obtained during the experiment are accurate.
D the temperature is equal in all tubes at the start of the experiment.

Questions 22, 23, 24 and 25 refer to the following information. Trypsin is a protein-digesting enzyme made in the human pancreas. Powdered milk suspension is a source of protein which has a white cloudy appearance. The accompanying diagram shows an investigation into the effect of boiling and into the effect of alkali on the activity of trypsin. This investigation consists of two experiments conducted at the same time.

	1	2	3	4
Tube contents	powdered milk + boiled trypsin + alkali	powdered milk + fresh trypsin + alkali	powdered milk + fresh trypsin + water	powdered milk + boiled trypsin + water
Appearance of contents at start	cloudy	cloudy	cloudy	cloudy
Appearance after 1 hour at 37 °C	cloudy	clear	slightly cloudy	cloudy

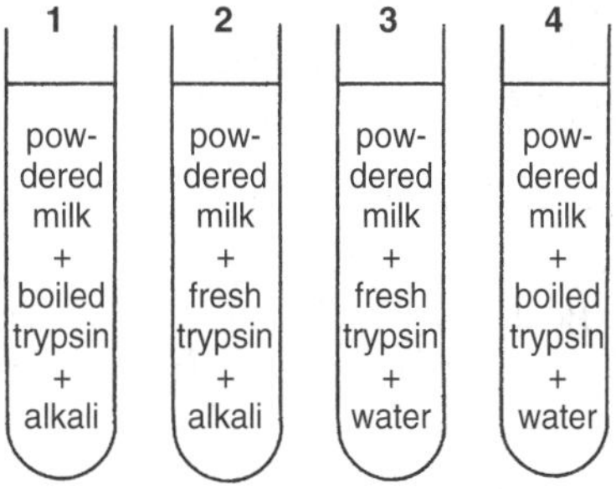

22 Which two tubes should be compared at the end of the experiment to draw a conclusion about the effect of boiling on trypsin's activity?

A 1 and 2 **B** 2 and 3 **C** 1 and 4 **D** 2 and 4

23 Which tube is the control in the experiment to investigate the effect of boiling on trypsin under alkaline conditions?

A 1 **B** 2 **C** 3 **D** 4

24 Which two tubes should be compared at the end of the experiment to draw a conclusion about the effect of alkali on trypsin's activity?

A 1 and 2 **B** 2 and 3 **C** 1 and 4 **D** 3 and 4

25 Which tube is the control in the experiment to investigate the effect of alkali on trypsin?

A 1 **B** 2 **C** 3 **D** 4

4 Respiration

Test I

1 Which of the following are BOTH raw materials used in the process of aerobic respiration?

A glucose and carbon dioxide
B carbon dioxide and water
C water and oxygen
D oxygen and glucose

2 The energy (in kJ) released by burning food can be calculated using the formula

$$\frac{4.2\,\mathbf{MT}}{1000}$$

where **M** = mass of water (g) and **T** = rise in temperature (°C).

Which of the following gives the number of kilojoules of energy released by the food shown in the accompanying diagram?

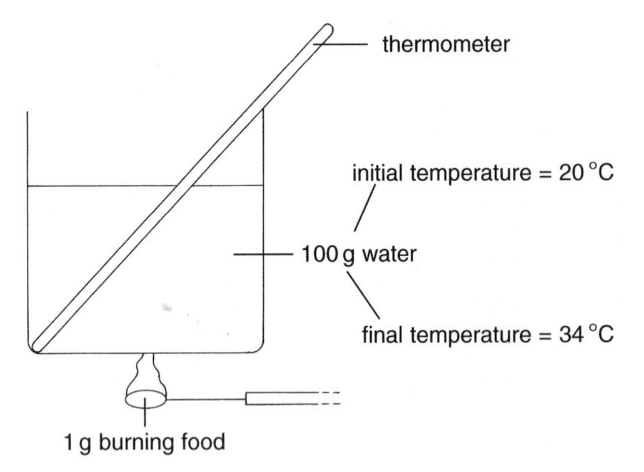

A 0.588 **B** 1.428
C 5.880 **D** 14.280

3 The following table refers to practices adopted during the use of a food calorimeter. Which line is CORRECT?

	Practice	Reason for adopting practice
A	thick lid fitted to top of water-filled chamber	to ensure even distribution of heat energy throughout the water
B	coiled copper chimney included in design	to ensure maximum transfer of heat to the water
C	circular stirrer moved up and down during burning of food	to ensure that food is completely burned
D	oxygen supplied during burning of food	to ensure that heat loss is reduced to a minimum

Items 4, 5, 6 and 7 refer to the accompanying tables. The first lists the energy content of several foods; the second shows the approximate energy requirements of different persons and age groups.

Food	Energy content (kJ/g)
almonds	24.3
apple pie	12.3
bread (white)	10.6
cheese (cheddar)	17.2
chocolate biscuits	20.8
cream (double)	18.8
fruit yoghurt	3.3
potatoes (boiled)	3.3
potatoes (chipped)	9.9
rice	15.0

Person	Approx. daily energy requirement (kJ)
infant	3000
6-year-old	7500
15-year-old girl	9500
15-year-old boy	12 000
woman (light work)	9500
woman (heavy work)	12 000
man (light work)	11 500
man (heavy work)	13 000
man (very heavy work)	16 000
adult patient in hospital	7500

4 Making potatoes into chips instead of boiling them increases their energy content by a factor of

A 0.33. **B** 3.00. **C** 3.30. **D** 6.60.

5 By how many times, on average, is the daily energy requirement of an adult patient in hospital greater than that of an infant?

A 2.5 **B** 3.0 **C** 4.5 **D** 7.5

6 Compared with his requirement at age 15 years, the daily energy requirement of a man doing very heavy work increases by

A 33.3% **B** 40.0% **C** 66.7% **D** 133.3%

7 A woman who has been doing light work changes to a job involving much heavy work. Her *additional* daily energy requirement could be met by eating

A 100 g of bread and 70 g of cheese.
B 100 g of apple pie and 50 g of double cream.
C 200 g of fruit yoghurt and 80 g of chocolate biscuits.
D 90 g of rice and 50 g of almonds.

8 Chocolate ice cream contains 9 kJ/g. Which of the following activities, when carried out for 30 minutes, uses the same amount of energy as is contained in 125 g of chocolate ice cream?

	Activity	Energy used (kJ/min)
A	playing football	37.5
B	walking up stairs	38.5
C	rowing	40.5
D	running	42.5

Items 9 and 10 refer to the accompanying diagram which shows an experiment set up to investigate the release of heat energy by respiring pea seeds. The table records the results over a period of one week.

9 Which set of results was obtained from flask P?

10 Which set of results was obtained from flask R?

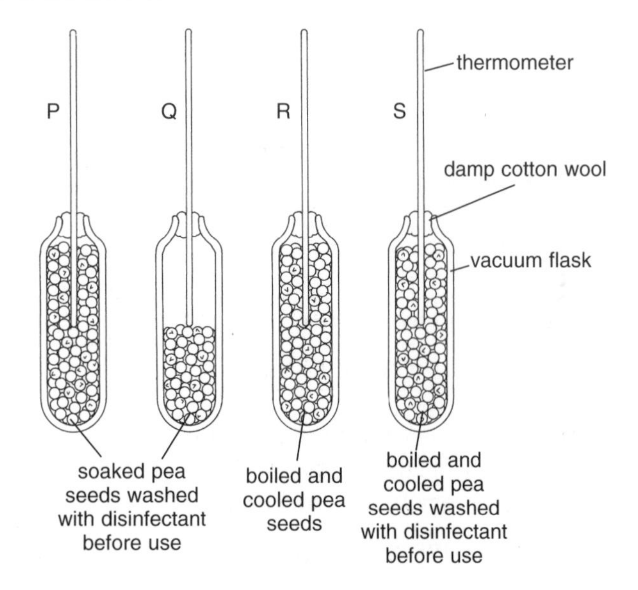

Set of results	Time (days)						
	1	2	3	4	5	6	7
A	20 °C	21 °C	22 °C	23 °C	23 °C	23 °C	23 °C
B	20 °C	22 °C	24 °C	26 °C	28 °C	28 °C	28 °C
C	20 °C	20 °C	20 °C	20 °C	20 °C	20 °C	20 °C
D	20 °C	20 °C	20 °C	21 °C	21 °C	22 °C	23 °C

11 Which of accompanying diagrams best represents the structure of a molecule of ATP (adenosine triphosphate)?

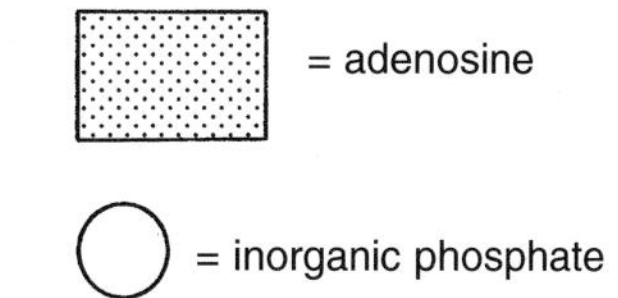

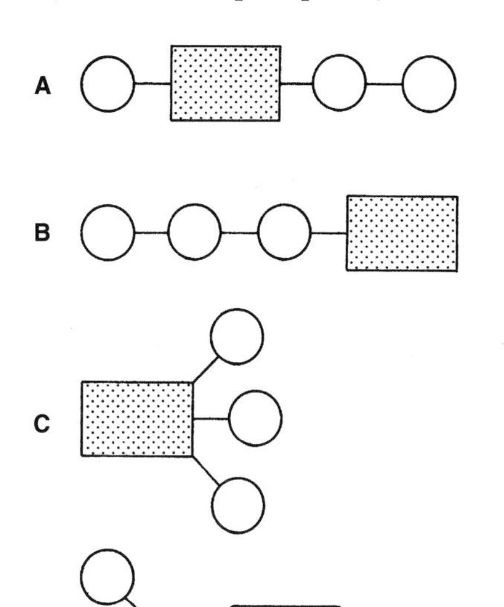

12 Which of the following equations represents the regeneration of ATP from its components?

A $ADP + Pi \xrightarrow{\text{energy taken in}} ATP$

B $ADP + Pi + Pi \xrightarrow{\text{energy taken in}} ATP$

C $ADP + Pi \xrightarrow{\text{energy released}} ATP$

D $ADP + Pi + Pi \xrightarrow{\text{energy released}} ATP$

13 As part of an investigation into the effect of different solutions on fresh muscle tissue, 12 drops of ATP were added to a strand of fresh muscle of initial length 50 mm. After a few minutes its length was found to be 42 mm.

Which line in the table correctly summarises the experiment?

	% difference in length of muscle strand	Reason for change
A	8	contraction of muscle fibres
B	8	relaxation of muscle fibres
C	16	contraction of muscle fibres
D	16	relaxation of muscle fibres

14 ATP is essential to every living cell because it

A makes energy instantly available when required.
B stores energy released during breakdown of ADP.
C speeds up the digestion of high energy foods.
D reacts with energy from glucose to form ADP.

15 The accompanying diagram shows the biochemical process of glycolysis.

Boxes X, Y and Z should contain the terms:

	X	Y	Z
A	glucose	4 ATP	lactic acid
B	pyruvic acid	2 ATP	glucose
C	lactic acid	4 ATP	pyruvic acid
D	glucose	2 ATP	pyruvic acid

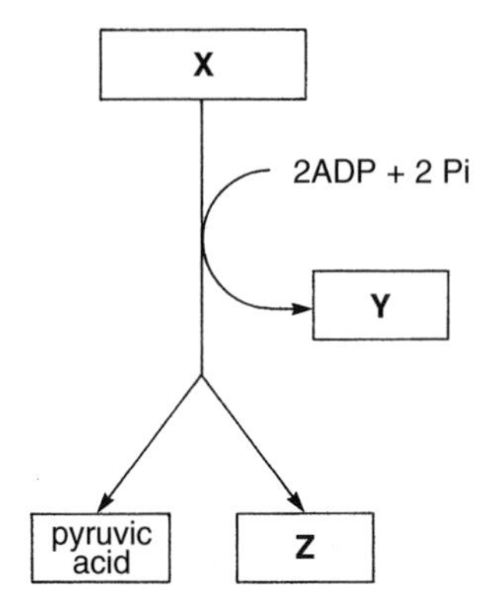

Questions 16 and 17 refer to the accompanying graph which shows the rate of oxygen uptake by a plant in darkness at different temperatures.

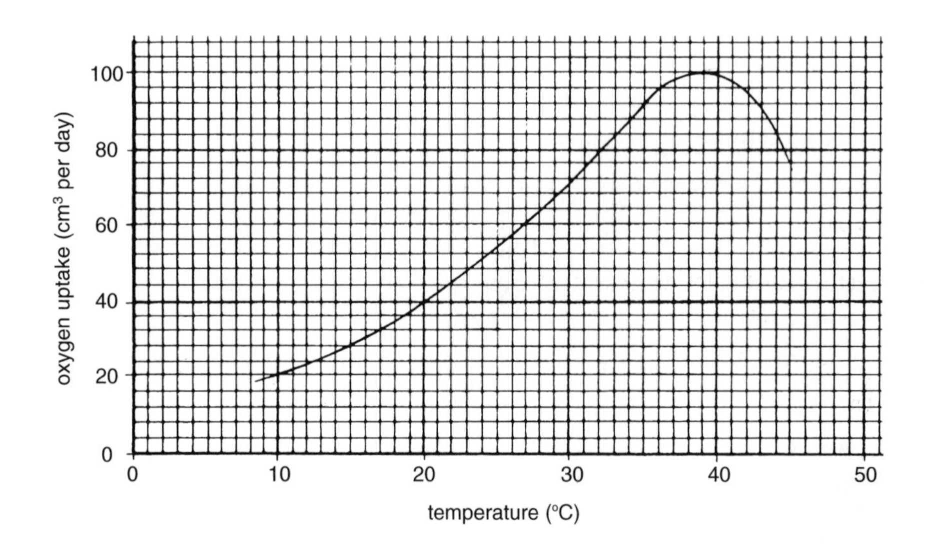

16 By how many times was the volume of oxygen taken up at 35 °C greater than that taken up at 10 °C?

A 3.5 **B** 4.3 **C** 4.6 **D** 72.0

17 What increase in temperature was required to double the volume of oxygen taken up at 24 °C?

A 10 **B** 15 **C** 25 **D** 50

Items 18 and 19 refer to the experiment shown in the accompanying diagram. It was set up to measure a grasshopper's rate of respiration.

After 30 minutes, the coloured liquid in the experiment was returned to its original level by depressing the syringe plunger from point X to point Y.

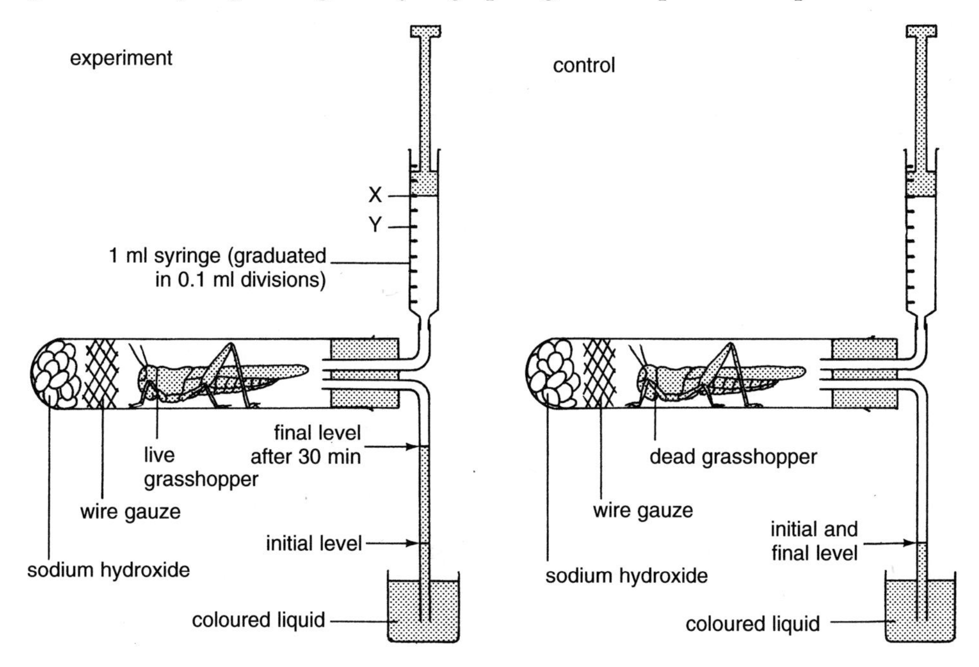

18 The rise in level of coloured liquid indicates that the

A grasshopper is taking in oxygen.
B sodium hydroxide is absorbing oxygen.
C grasshopper is giving out carbon dioxide.
D sodium hydroxide is releasing carbon dioxide.

19 From this experiment it can be concluded that the grasshopper's rate of

A carbon dioxide output is 0.2 ml/hour.
B carbon dioxide output is 4.0 ml/hour.
C oxygen consumption is 2.0 ml/hour.
D oxygen consumption is 0.4 ml/hour.

20 As part of an investigation into aerobic respiration by germinating barley grains, the flasks shown in the accompanying diagram were set up and kept at 24 °C.

Bicarbonate indicator changes colour from red to yellow in the presence of a relatively high concentration of carbon dioxide. In which flask did the indicator change colour first?

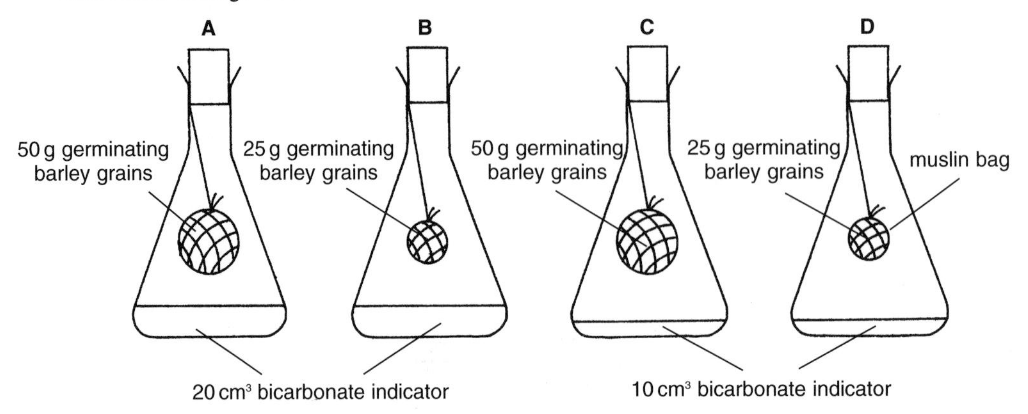

Items 21 and 22 refer to the accompanying diagram which represents the process of anaerobic respiration in plant cells.

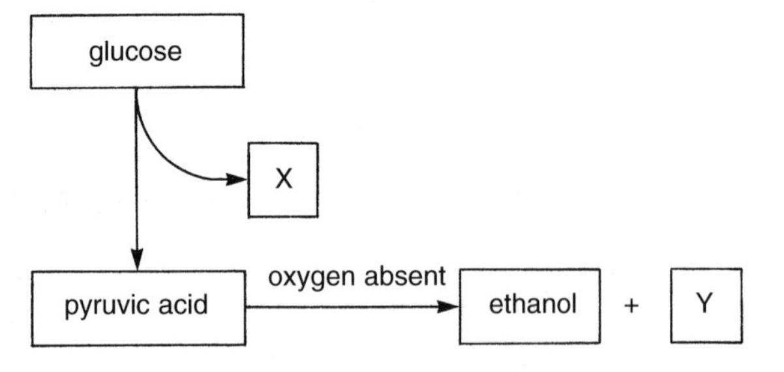

21 The number of molecules of ATP formed at position X following the breakdown of one molecule of glucose is

A 2. **B** 18. **C** 36. **D** 38.

22 The substance released at position Y is

A water. **B** oxygen.
C hydrogen. **D** carbon dioxide.

Questions 23, 24 and 25 refer to the accompanying graph which shows the effect of a period of exercise followed by a period of rest on the lactic acid concentration of the blood of a healthy fit teenager.

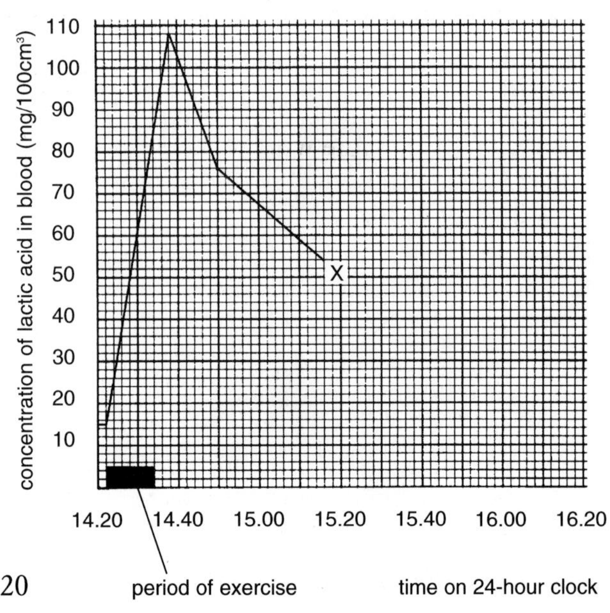

23 For how many minutes did the period of exercise last?

A 6 **B** 10 **C** 12 **D** 14

24 How many minutes did it take for the concentration of lactic acid to drop from its highest level to 50% of its highest level?

A 19 **B** 20 **C** 38 **D** 40

25 If the trend at X continues, at what time will the initial level of lactic acid in the blood be reached?

A 15.56 **B** 16.02 **C** 16.08 **D** 16.20

Test 2

Items 1, 2, 3 and 4 refer to the following experiments where, in each case, an attempt is made to measure the number of kilojoules (kJ) given out by 1 gram of food.

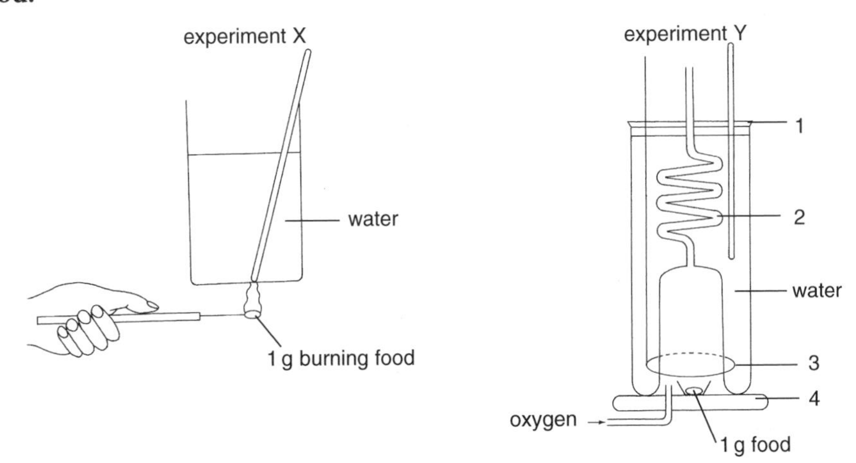

1 The apparatus used in experiment Y is called a

A calorie. **B** centrifuge. **C** calorimeter. **D** calibrator.

2 Which of the following has been omitted from the diagram of experiment Y?

A chimney **B** thermometer **C** stirrer **D** igniter

3 Which of the numbered structures in the diagram of experiment Y does NOT help to prevent the escape of heat energy?

A 1 **B** 2 **C** 3 **D** 4

4 A possible source of error in experiment X is that the heat energy is not evenly spread throughout the water. This source of error is overcome in experiment Y by structures

A 1 and 2. **B** 2 and 3. **C** 3 and 4. **D** 1 and 3.

Items 5 and 6 refer to the following information. The results in the accompanying table were obtained using apparatus Y shown in the first diagram in this test. (4.2 kJ = amount of energy required to raise the temperature of 1000 g of water by 1 °C.)

5 When burned, 1 g of one of the foods in the table was found to raise the temperature of 1000 g of water by 0.5 °C. Identify the food.

A peach **B** parsnip
C sweetcorn **D** trifle

6 How many grams of trifle would have to be burned to raise the temperature of 1000 g of water by 2 °C?

A 1 **B** 2 **C** 3 **D** 4

Food	Number of kilojoules (kJ) released on burning 1 g of food
peach	1.5
parsnip	2.1
sweetcorn	4.2
trifle	8.4

Items 7 and 8 refer to the data in the following table which were obtained from a series of experiments involving a group of 18-year-olds.

Activity	Average rate of oxygen consumption (cm^3/s)	Average rate of energy consumption (kJ/min)
cycling	21	26
swimming	23	28
skiing	26	33
brick-laying	29	36

7 From the data it can be concluded that the average rate of oxygen consumption is

A inversely related to CO_2 output.
B greatest for sporting activities.
C directly related to energy consumption.
D inversely related to energy released as heat.

8 Which pair of bars in the accompanying graph represent skiing?

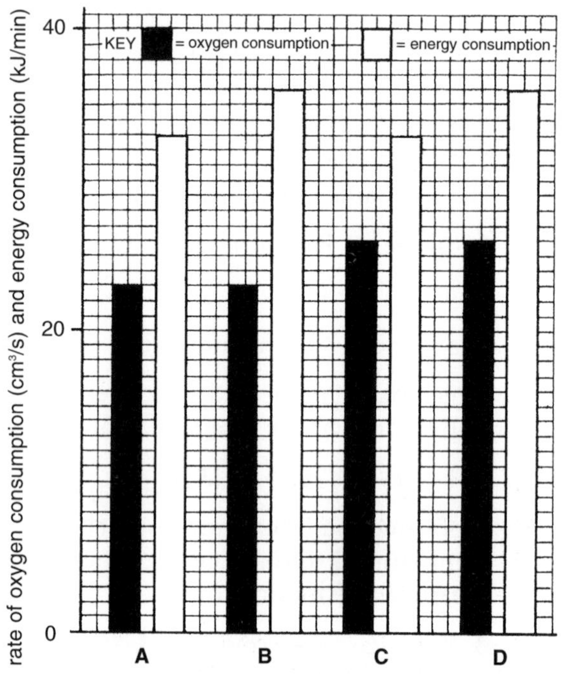

9 During jogging, a 16-year-old was found to use 40 kJ per minute. The energy for 15 minutes of this activity could be supplied by consuming 50 g of one of the foods in the accompanying table. Which one?

	Food	Energy content (kJ/g)
A	honey	12
B	sucrose	19
C	biscuit	21
D	chocolate	24

10 Which of the following would NOT affect the daily energy requirement of a 5-year-old child?

A the climate in which the child lives
B the height of the child
C the weight of the child
D the sex of the child

11 Three vacuum flasks were set up (as shown in the first of the accompanying diagrams) to investigate the release of heat energy by respiring pea seeds. The results were graphed as shown in the second diagram.

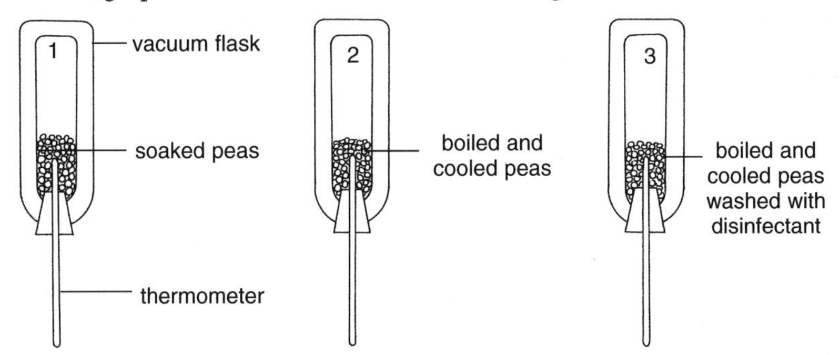

Which of the following correctly matches each flask with the line graph of its results?

A X=2, Y=1, Z=3
B X=2, Y=3, Z=1
C X=1, Y=2, Z=3
D X=3, Y=1, Z=2

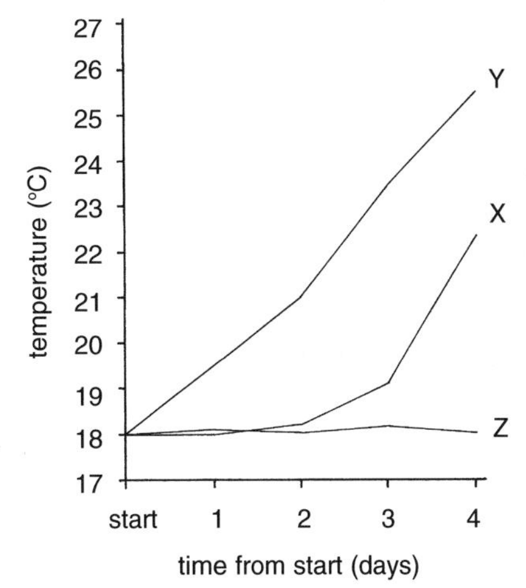

Items 12 and 13 refer to the following information. During an inspection, a pest control officer recorded the temperature of the grain in a grain store at different depths below the surface (without disturbing the grain).

The accompanying graph shows his results. Four populations of insect pest (A, B, C and D) were found at the points indicated on the graph.

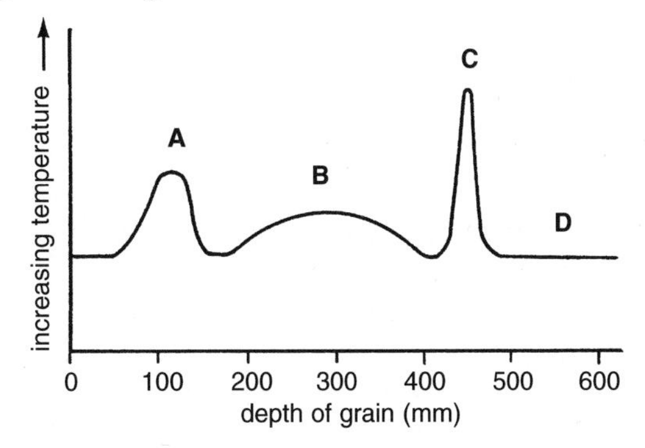

12 In which population were the insects dead?

13 Which insect population was most widespread in its vertical distribution in the grain?

14 The accompanying diagram illustrates the process of tissue respiration. Which box represents the correct position of ATP in this scheme?

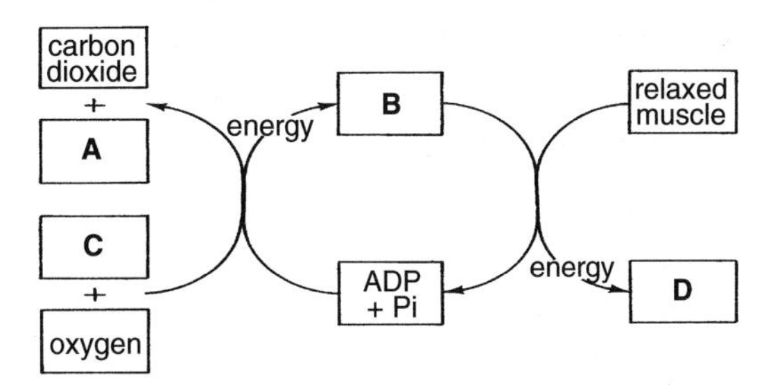

15 The energy released during the breakdown of ATP is NOT used for

A synthesis of protein.
B transmission of nerve impulses.
C production of enzymes.
D digestion of fats.

16 Which of the following equations correctly represents muscular action?

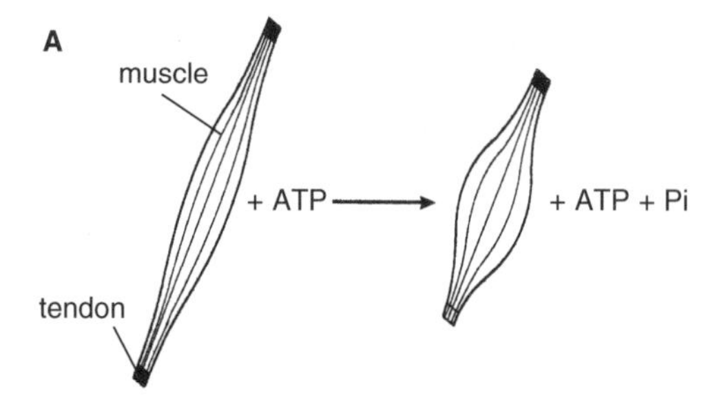

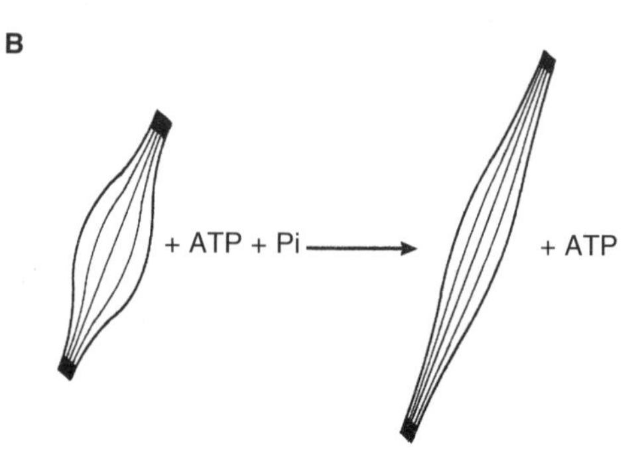

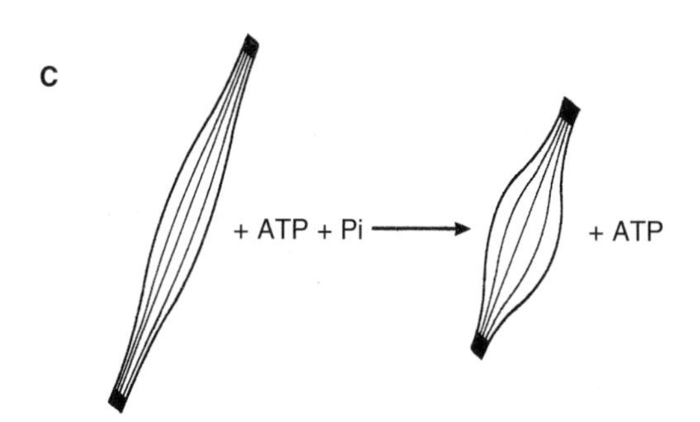

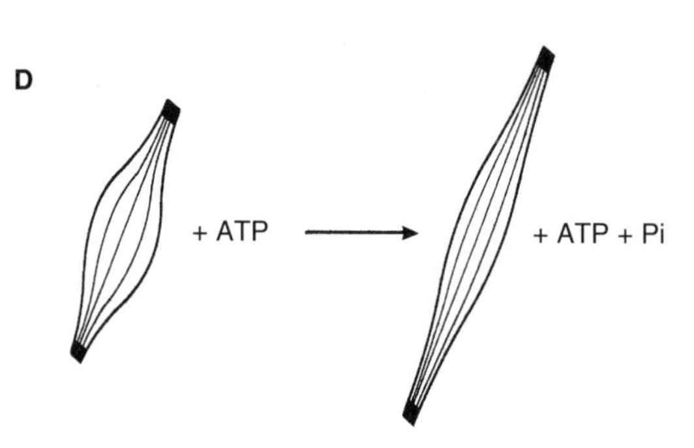

17 The breakdown of glucose to pyruvic acid is called

A glycolysis. **B** glycogenesis.
C dephosphorylation. **D** aerobic respiration.

Questions 18 and 19 refer to the accompanying diagram of the aerobic breakdown of pyruvic acid. They also refer to the following possible answers.

A glucose
B oxygen
C lactic acid
D carbon dioxide

18 Which substance is released at positions X, Y and Z?

19 Which substance enters the system at point Q?

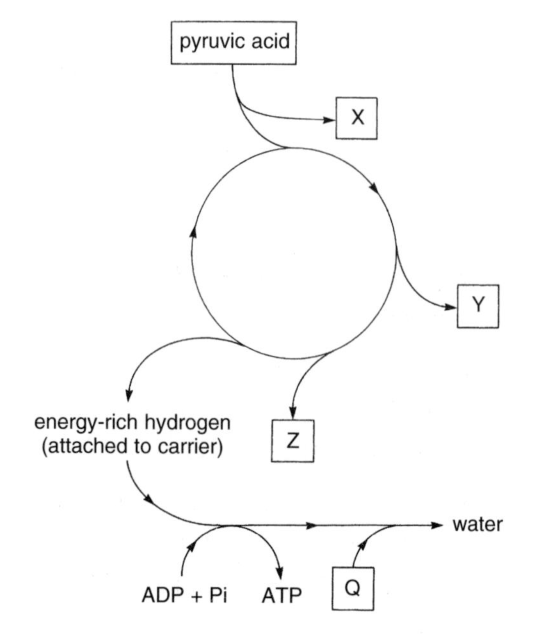

20 Which of the following equations represents the aerobic breakdown of one molecule of glucose?

A glucose + carbon dioxide + 38 ATP → water + oxygen + 38 ADP + 38 Pi
B carbon dioxide + glucose + 38 ADP + 38 Pi → oxygen + water + 38 ATP
C oxygen + glucose + 38 ADP + 38 Pi → carbon dioxide + water + 38 ATP
D glucose + oxygen + 38 ATP → water + carbon dioxide + 38 ADP + 38 Pi

21 If an animal of mass 100 g consumes 50 cm^3 of oxygen in 30 minutes, then its respiratory rate in cm^3 of oxygen used per gram of body tissue per hour is

A 0.25. **B** 0.50. **C** 1.00. **D** 2.00.

Questions 22 and 23 refer to the following possible answers.

A water **B** glucose **C** alcohol **D** lactic acid

22 Which substance is required by a yeast cell for anaerobic respiration?

23 Which substance would be produced by skeletal muscle fibres respiring in the absence of oxygen?

Items 24 and 25 refer to the accompanying diagram which shows an experiment set up to investigate the effect of temperature on rate of anaerobic respiration by yeast cells. The table gives the results obtained.

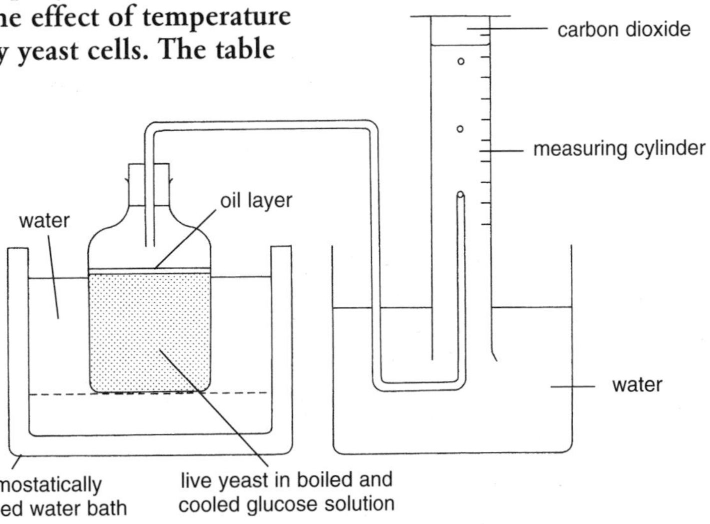

carbon dioxide
measuring cylinder
oil layer
water
water
thermostatically controlled water bath
live yeast in boiled and cooled glucose solution

	Temperature (°C)					
	10	15	20	30	35	40
Average volume of CO_2 released per hour (cm^3)	4	11	16	34	43	45

24 If the experiment were repeated at 25 °C, the most likely volume (cm^3) of CO_2 released per hour would be

A 14. **B** 26. **C** 33. **D** 43.

25 If the experiment were repeated at 70 °C, the most likely volume (cm^3) of CO_2 released per hour would be

A 0. **B** 12. **C** 45. **D** 59.

5 Photosynthesis

Test 1

1 Which of the following word equations correctly represents the process of photosynthesis?

A water + carbon dioxide → sugar + oxygen + energy
B carbon dioxide + water + energy → sugar + oxygen
C sugar + carbon dioxide → water + oxygen + energy
D carbon dioxide + sugar + energy → water + oxygen

Items 2 and 3 refer to the following possible answers.

A starch **B** cellulose
C chloroplast **D** chlorophyll

2 Which term refers to a green, discus-shaped structure found in a leaf cell?

3 Which term refers to the green pigment that traps light energy?

Items 4 and 5 refer to the accompanying diagram of a transverse section through part of a green leaf.

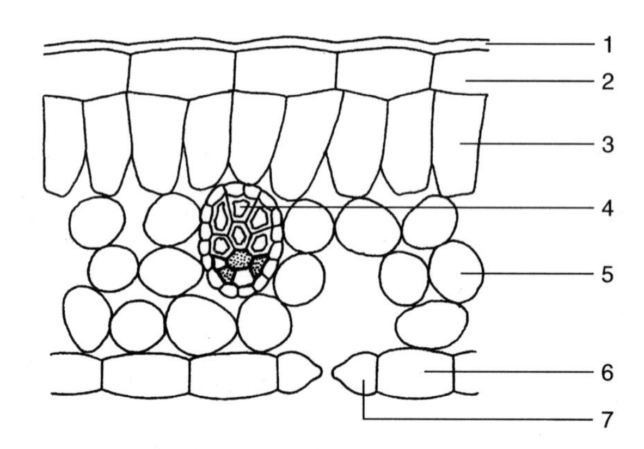

4 In which region is most light energy trapped?

A 1 **B** 2 **C** 3 **D** 5

5 In which of the following pairs does photosynthesis occur in NEITHER region?

A 2 and 5 **B** 5 and 6
C 2 and 7 **D** 4 and 6

Items 6 and 7 refer to the accompanying diagram which shows some of the steps carried out to test a leaf for the presence of starch.

6 The correct sequence of the steps is

A Y, W, Z, X.
B Z, W, Y, X.
C Y, W, X, Z.
D Z, Y, W, X.

7 The reason for carrying out step W is to

A kill the leaf cells.
B soften the leaf discs.
C remove chlorophyll from the leaf cells.
D extract oxygen bubbles from the leaf discs.

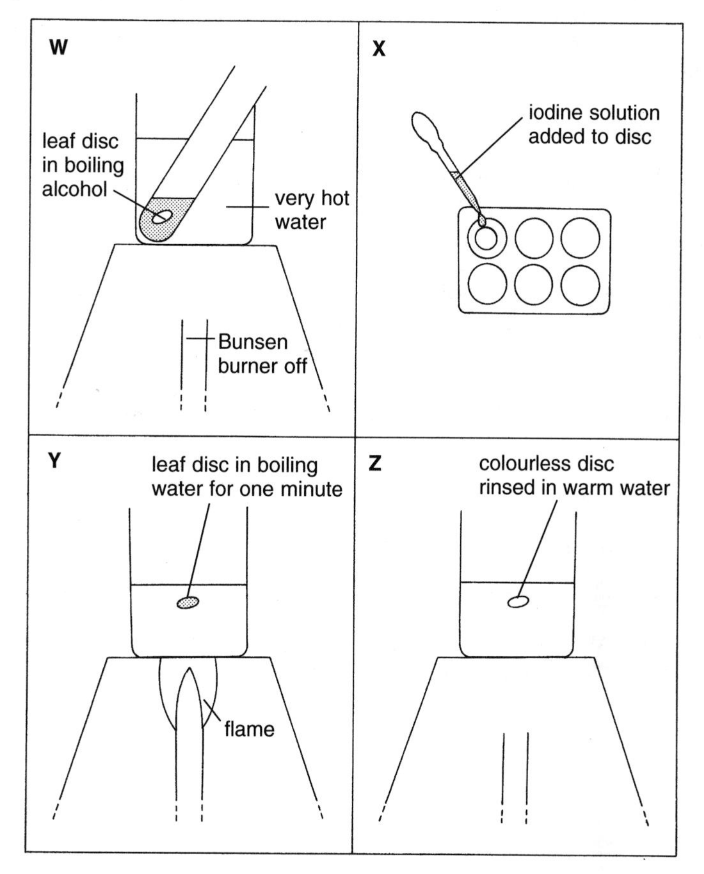

8 The experiment shown in the accompanying diagram was set up to investigate whether light is necessary for photosynthesis.

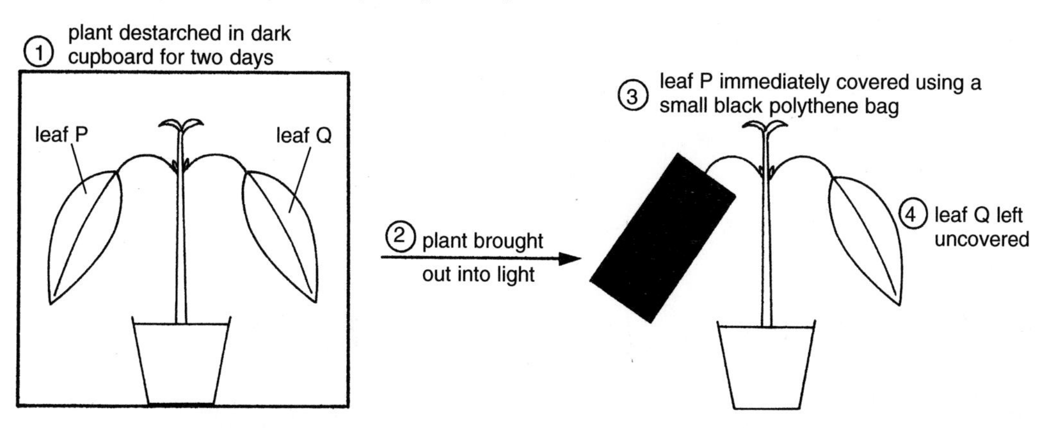

The validity of this experiment could have been increased by

A leaving leaf P uncovered.
B covering leaf Q with a black polythene bag.
C covering leaf Q with a transparent polythene bag.
D returning the whole plant to the dark cupboard.

Items 9 and 10 refer to the experiment shown in the accompanying diagram. The plant was left in sunlight for two days and then leaf discs W, X, Y and Z were tested for the presence of starch.

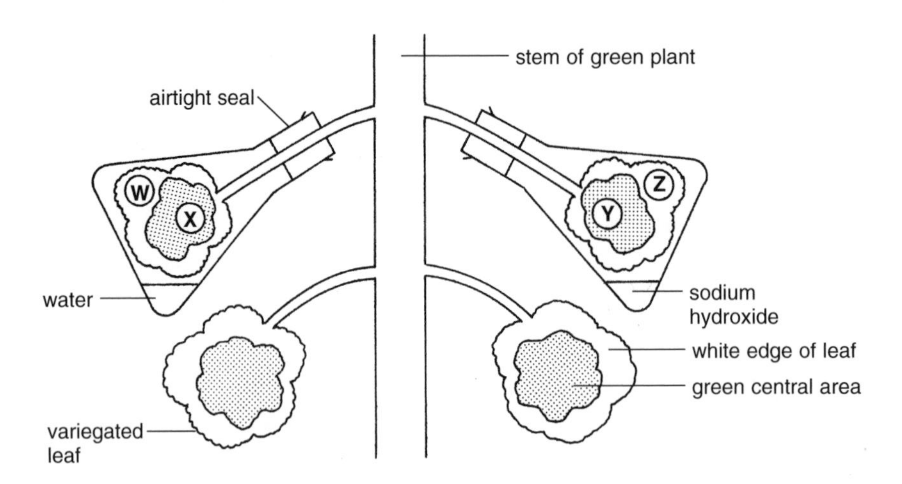

9 The leaf disc found to contain starch was

A W. **B** X. **C** Y. **D** Z.

10 This experiment proves that in order to photosynthesise, a plant must have

A sunlight and carbon dioxide.
B carbon dioxide and water.
C chlorophyll and sunlight.
D carbon dioxide and chlorophyll.

11 A variegated leaf on a destarched ivy plant was treated as shown in the accompanying diagram.

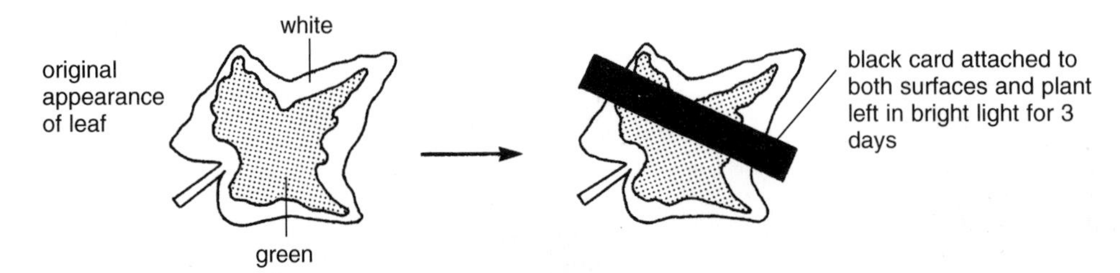

Which of the following shows the appearance of the leaf after testing it for starch?

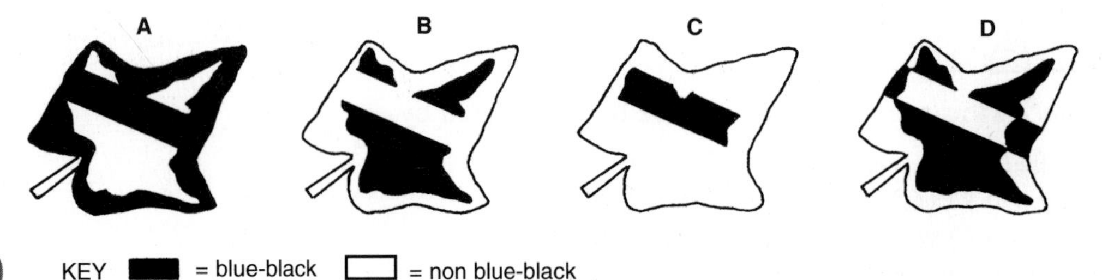

12 The total volume of carbon dioxide daily entering the plant shown in the accompanying diagram is 24 000 mm^3.

The daily rate of diffusion of carbon dioxide into the plant in mm^3 CO_2 per mm^2 of leaf is

A 0.16. **B** 6.00.
C 48.00. **D** 3000.00.

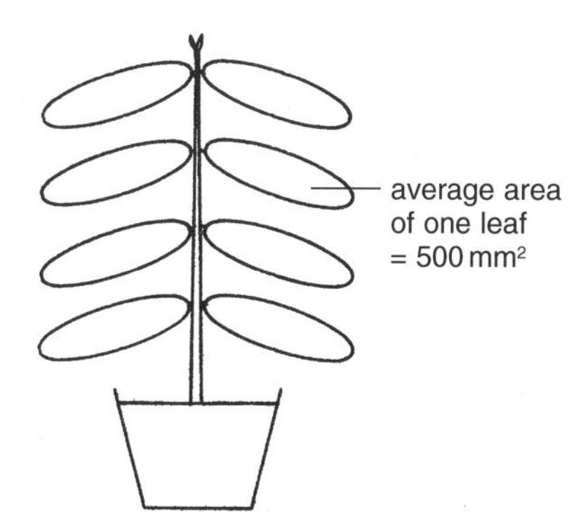

Questions 13 and 14 refer to the accompanying diagram which shows the process of photolysis. The two questions also refer to the following possible answers.

A water **B** oxygen
C hydrogen **D** carbon dioxide

13 What is the correct identity of chemical substance X?

14 What is the correct identity of chemical substance Y?

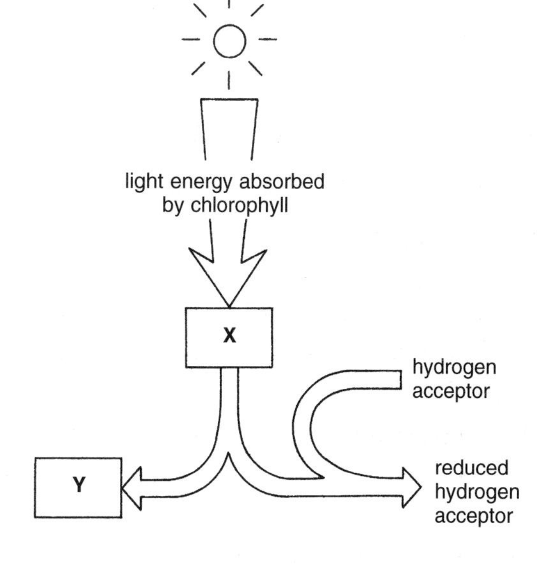

Items 15 and 16 refer to the accompanying diagram of carbon fixation.

15 This cyclical chemical reaction is dependent upon the presence of BOTH

A enzymes and a suitable temperature.
B a suitable temperature and light energy.
C light energy and glucose. ✗
D glucose and enzymes.

16 The correct identity of substance Z is

A water. **B** oxygen.
C simple sugar. **D** carbon dioxide.

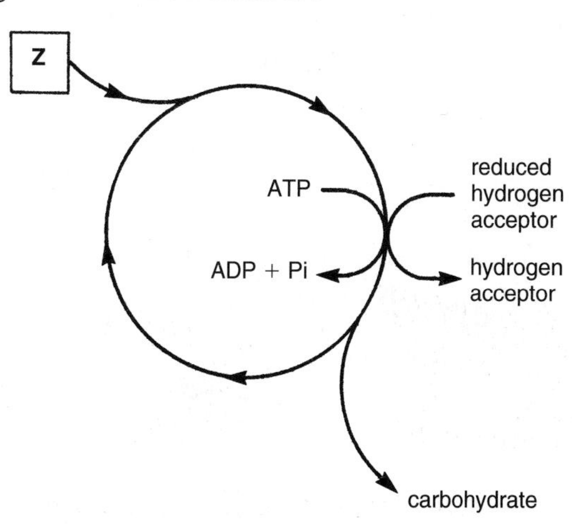

17 Excess sugar formed by photosynthesis may be converted into other carbohydrates. Which line in the following table is CORRECT?

	Storage carbohydrate	Structural carbohydrate
A	cellulose	starch
B	glucose	glycogen
C	starch	cellulose
D	glycogen	glucose

18 Which of the following CANNOT be used to measure rate of photosynthesis?

A volume of oxygen released per unit time
B volume of carbon dioxide taken up per unit time
C mass of carbohydrate produced per unit time
D volume of water vapour released per unit time

19 The graph in the accompanying diagram shows the results from a photosynthesis experiment.

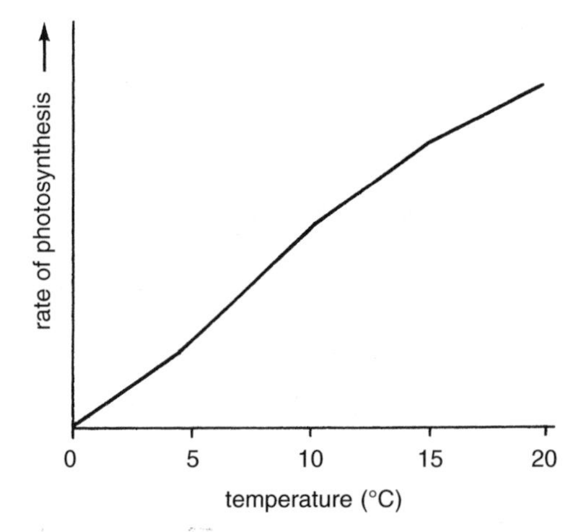

Which of the following pairs of environmental factors must be kept constant during this experiment to make it valid?

A light intensity and temperature.
B temperature and carbon dioxide concentration
C water content and oxygen concentration
D light intensity and carbon dioxide concentration

Questions 20 and 21 refer to the accompanying graph which shows the effect of increasing light intensity on rate of photosynthesis by Canadian pondweed.

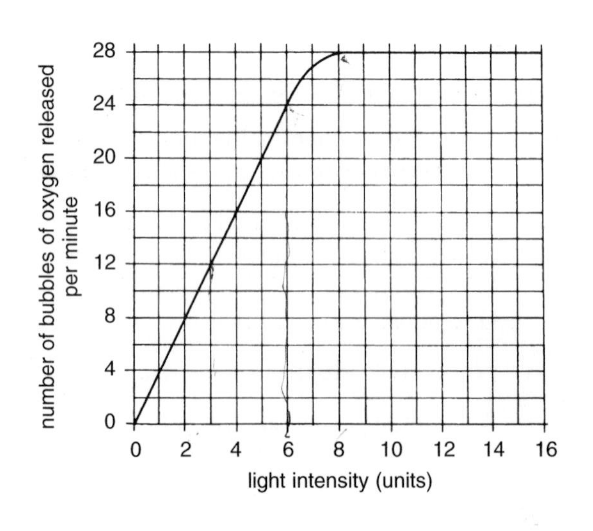

20 By how many times was the rate of photosynthesis at 8 units of light greater than that at 2 units of light?

A 3.5 **B** 4.0 **C** 20.0 **D** 224.0

21 What increase in number of units of light intensity was required to double the rate of photosynthesis occurring at 3 units of light intensity?

A 3 **B** 6 **C** 12 **D** 24

Items 22 and 23 refer to the graphs in the accompanying diagram which show the effects of several factors on the rate of photosynthesis by a green plant. These two items also refer to the following possible answers.

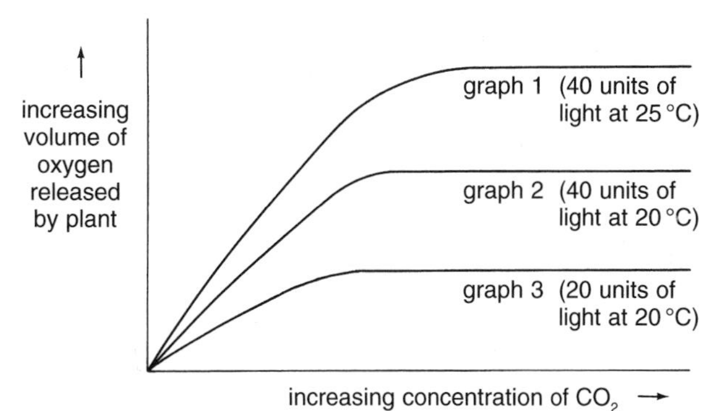

A temperature
B light intensity
C volume of oxygen
D carbon dioxide concentration

22 Which factor caused graph 3 to level off?

23 Which factor caused graph 2 to level off?

24 Early crops of tomatoes produced in greenhouses are often given extra carbon dioxide since the normal atmospheric concentration (0.03%) tends to act as a limiting factor. The additional carbon dioxide is provided

A during the night only since plants respire at night.
B during the day only since photosynthesis occurs in light.
C during the day and night since plants respire 24 hours a day.
D on cloudy days only when light is a limiting factor.

25 The accompanying tables refer to a survey done on the production of a certain crop plant in Scotland.

Month seeds were sown	Average mass of crop produced (tonnes/hectare)
March	30
April	35
May	26

Number of plants grown per hectare	Average mass of crops produced (tonnes/hectare)
60 000	31
80 000	42
100 000	33

Which combination of planting conditions would give the best crop yield?

	Month seeds are sown	Number of plants planted per hectare
A	April	60 000
B	March	80 000
C	April	80 000
D	May	100 000

Test 2

Items 1 and 2 refer to the following possible answers.

A water and oxygen　　**B** oxygen and glucose
C glucose and carbon dioxide　　**D** carbon dioxide and water

1 Which substances are BOTH raw materials used in the process of photosynthesis?

2 Which substances are BOTH products which result from the process of photosynthesis?

3 Which of the following statements is CORRECT? Chlorophyll is found in

A vacuoles in all plant cells.
B vacuoles in green leaf cells.
C chloroplasts in all plant cells.
D chloroplasts in green leaf cells.

4 The accompanying diagram shows part of a green leaf in transverse section.

In which of the following pairs does photosynthesis occur in BOTH regions?

A Q and S　　**B** P and R
C R and S　　**D** T and U

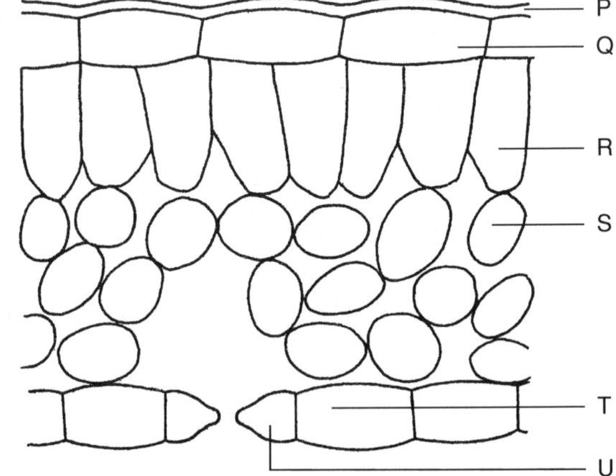

5 During photosynthesis

A light energy is converted to chemical energy contained first in ATP and later in glucose.
B light energy is converted to chemical energy contained first in glucose and later in ATP.
C chemical energy is converted to light energy contained first in ATP and later in glucose.
D chemical energy is converted to light energy contained first in glucose and later in ATP.

6 The following list gives the steps carried out when testing a leaf for the presence of starch.
1 decolourise
2 boil in water
3 add iodine solution
4 rinse in warm water
The correct sequence of these steps is

A 2, 3, 1, 4.　**B** 1, 2, 3, 4.　**C** 2, 1, 4, 3.　**D** 1, 2, 4, 3.

Items 7, 8 and 9 refer to the accompanying diagram which shows an experiment set up to investigate whether light is necessary for photosynthesis.

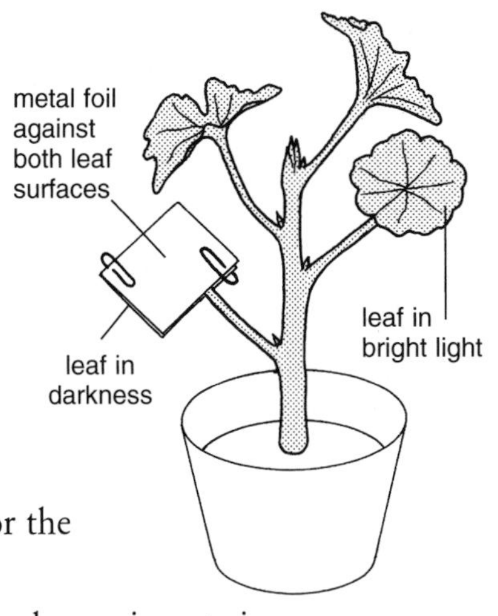

7 Before being used in the experiment, the plant should be

A destarched.
B brightly illuminated.
C exposed to sodium hydroxide.
D given extra carbon dioxide.

Items 8 and 9 also refer to the possible answers illustrated in the second diagram.

8 Which of these set-ups would act as an effective control for the original experiment?

9 Which of these set-ups is an improved version of the original experiment since it eliminates a source of error?

Questions 10, 11 and 12 refer to the accompanying diagram which shows experiments set up to investigate which factors are necessary for photosynthesis.

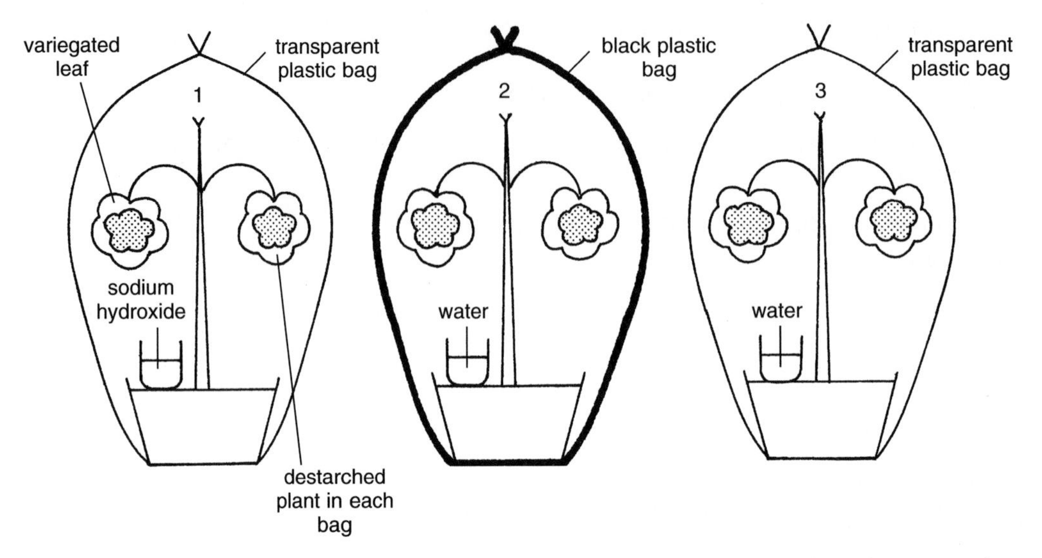

The three questions also refer to the possible answers illustrated in the second diagram.

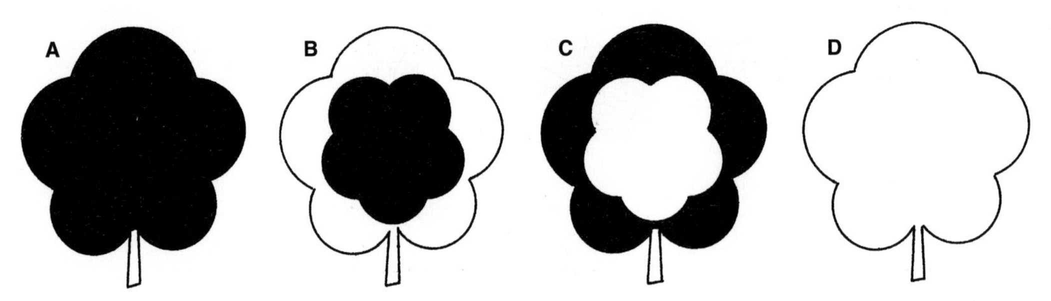

10 Which part of the diagram shows a leaf from plant 1 tested with iodine solution after two days?

11 Which part of the diagram shows a leaf from plant 2 tested with iodine solution after two days?

12 Which part of the diagram shows a leaf from plant 3 tested with iodine solution after two days?

13 The average intensity of solar radiation falling on a crop of sugar cane plants was found to be 16 000 kJ/m²/day. If 4.5% of this energy was absorbed by the plants during photosynthesis, the number of kilojoules trapped per m² of plants per hour would be

A 30. **B** 720. **C** 3000. **D** 17 280.

14 The oxygen released by a green plant as a result of photolysis comes from

A air. **B** glucose.
C water. **D** carbon dioxide.

15 The light-dependent stage of photosynthesis results in the formation of two compounds needed for the carbon fixation stage. These are

A ATP and reduced hydrogen acceptor.
B reduced hydrogen acceptor and ADP.
C ADP and inorganic phosphate.
D reduced hydrogen acceptor and inorganic phosphate.

16 Which of the answers given below CORRECTLY refers to the following three statements about carbohydrates formed by a plant?

1 Glucose is used as a source of energy.
2 Cellulose is used to build cell walls.
3 Excess sugar is stored as starch.

A 1 and 2 only are correct
B 1 and 3 only are correct
C 2 and 3 only are correct
D 1, 2 and 3 are correct

17 The accompanying diagram shows a potato plant and a close-up of two of its cells. Which line in the table that follows is CORRECT?

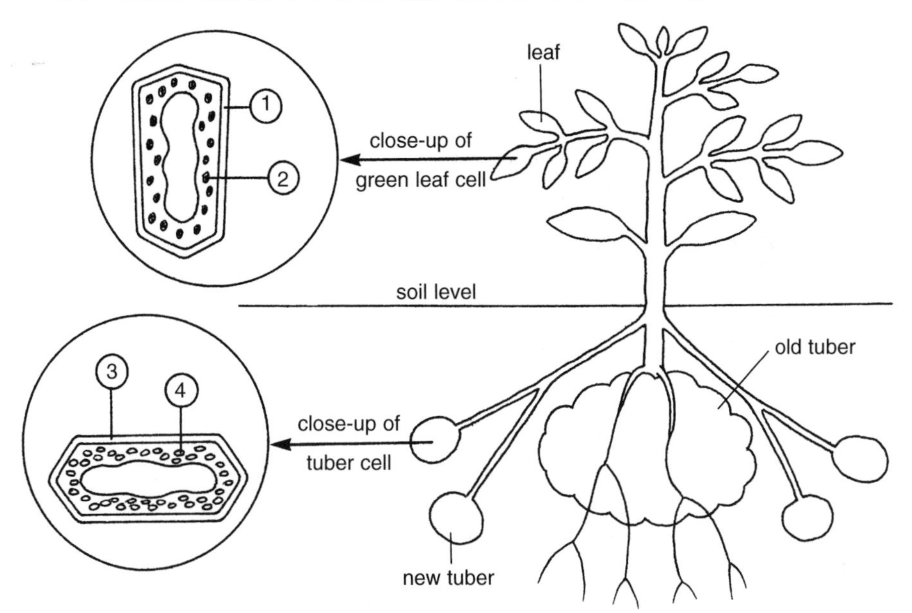

	Type of carbohydrate present at location:							
	1		2		3		4	
	S	C	S	C	S	C	S	C
A	✗	✓	✓	✗	✗	✓	✓	✗
B	✓	✗	✗	✓	✓	✗	✗	✓
C	✓	✗	✓	✗	✗	✓	✗	✓
D	✗	✓	✗	✓	✓	✗	✓	✗

S = starch C = cellulose ✓ = present ✗ = absent

18 The accompanying diagram shows the apparatus assembled by a group of students to investigate the effect of light intensity on rate of photosynthesis.

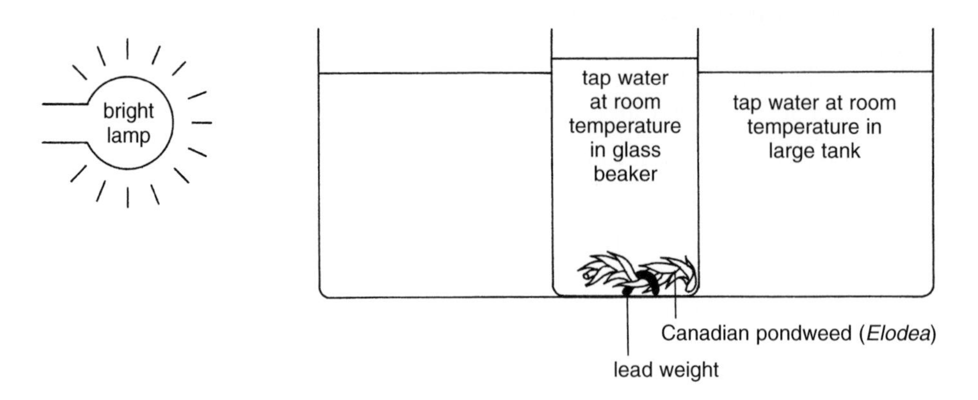

Which of the following should be done to improve the procedure and eliminate a possible limiting factor?

A Insert a transparent heat shield between the lamp and the plant.
B Immerse the green plant in pond water rich in carbon dioxide.
C Use dilute sodium hydrogen carbonate solution in the large tank.
D Remove the lead weight to allow the plant to float at the surface of the water.

19 Which of the following graphs most accurately represents the effect of increasing temperature on rate of photosynthesis in *Elodea* waterweed?

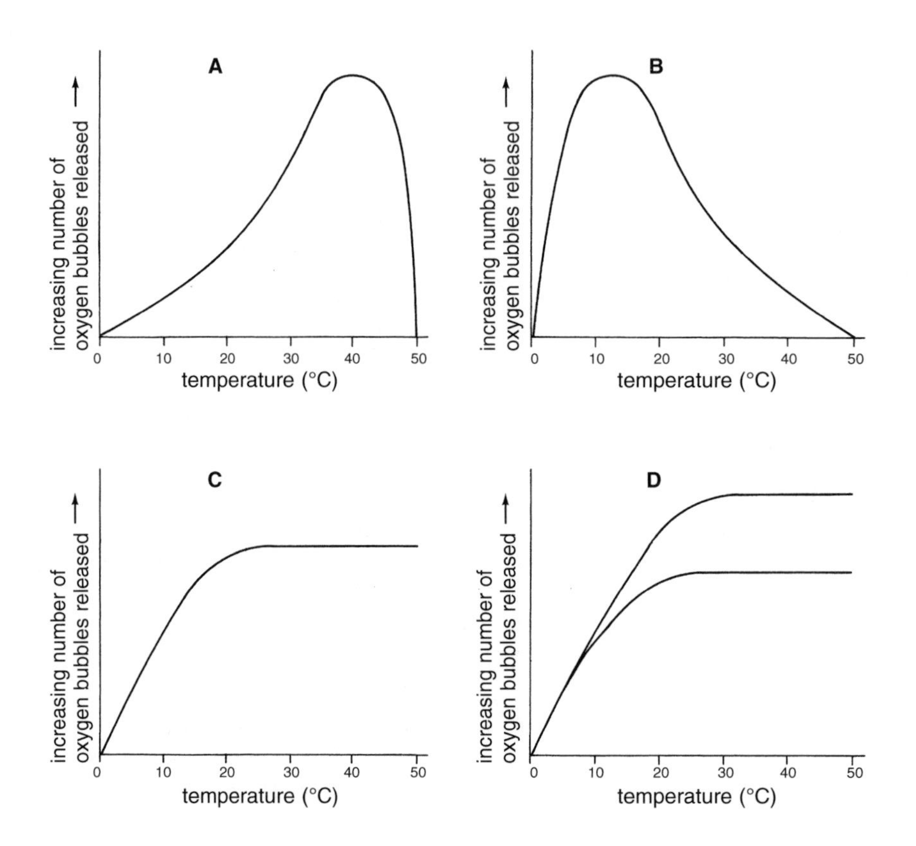

20 Assume that a plant must receive 3 units of CO_2, 3 units of water and 6 units of light energy to make 1 unit of sugar by photosynthesis.

Which of the plants in the following table would be able to make the greatest number of sugar units?

	Units of CO_2 available to plant	**Units of water available to plant**	**Units of light available to plant**
A	12	12	48
B	18	18	36
C	18	36	24
D	36	18	18

21 The graph in the accompanying diagram shows the effect of varying the CO_2 concentration at two different light intensities on the rate of photosynthesis by *Elodea*.

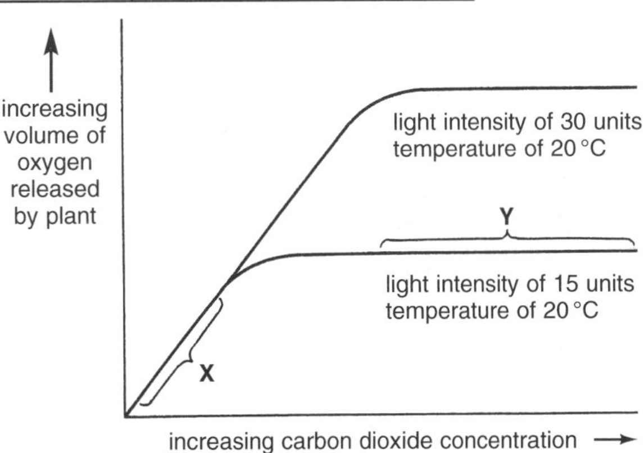

Which factors are limiting the rate of photosynthesis at regions X and Y on the graph?

	X	**Y**
A	light intensity	CO_2 concentration
B	light intensity	temperature
C	CO_2 concentration	light intensity
D	CO_2 concentration	temperature

Items 22 and 23 refer to the accompanying graph which shows the effect of light intensity on the rate of photosynthesis at three different concentrations of carbon dioxide.

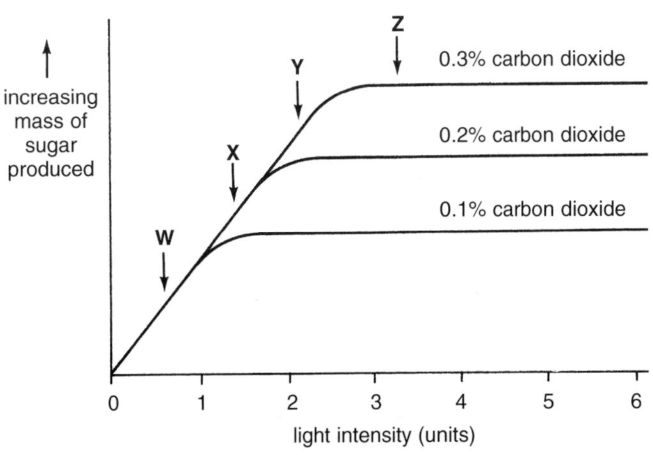

22 Which of the following light intensities was the limiting factor at all three concentrations of carbon dioxide?

A 0–1 units **B** 1–2 units **C** 2–3 units **D** 3–4 units

23 Carbon dioxide concentration was the limiting factor at

A point W only. **B** point Z only.
C points X and Y only. **D** points W, X, Y and Z.

24 The accompanying diagram shows four greenhouses set up to grow tomato plants. (The plants are not shown.)

In which greenhouse is CO_2 concentration definitely the factor limiting photosynthesis?

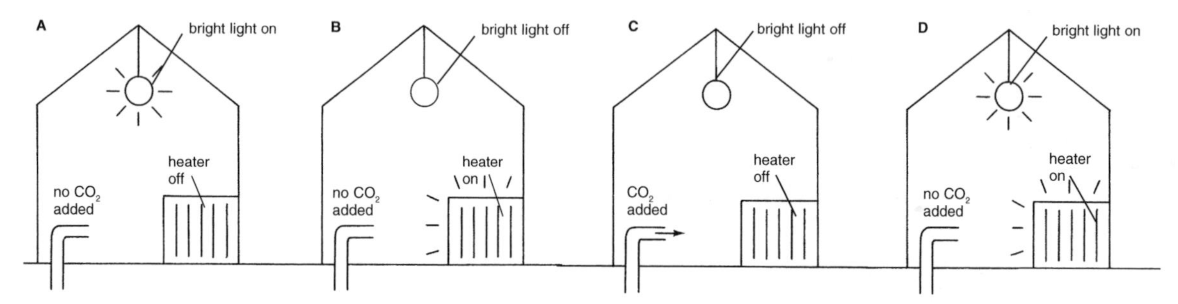

25 The table below shows the results of a survey on mass of food produced by a variety of plants as a result of photosynthesis. Which type of plant matches the following description?

This plant is a form of natural vegetation whose daily rate of photosynthesis during its season of most rapid growth is less than that of maize yet its annual rate of food production exceeds that of maize.

	Type of plant	Mass of carbon fixed during season of most rapid growth (g/day)	Mass of organic matter produced (tonnes/hectare/year)
	maize	8	25
A	spring wheat	10	22
B	swamp moss	7	23
C	seaweed	4	32
D	Scots pine	3	18

6 Energy flow

Test 1

Items 1, 2 and 3 refer to the following possible answers.

A community **B** ecosystem
C population **D** habitat

1 Which term refers to a natural biological unit composed of living and non-living parts?

2 Which term refers to all of the organisms that live together in an area?

3 Which term refers to the place where an organism lives?

4 Which of the following refers to a population of organisms?

A all of the reptiles in a desert
B all of the mammals in a jungle
C all of the hermit crabs in a rock pool
D all of the flowing plants in a woodland

Questions 5, 6 and 7 refer to the following possible answers.

A green plant that makes food by photosynthesis
B animal that hunts other animals for its food supply
C micro-organism that obtains its energy by breaking down dead organic material
D organism that requires a ready-made food supply of plant material

5 Which description fits a primary consumer?

6 Which answer describes a predator?

7 Which description refers to a decomposer?

8 The accompanying stick graph shows the temperature changes that took place in a compost heap over a period of 24 days.

From this graph it can be concluded that the decay microbes

A increased in number from day 1 to day 10 only.
B increased in number from day 1 to day 20.
C gave out heat energy from day 10 to day 20 only.
D gave out heat energy from day 1 to day 20.

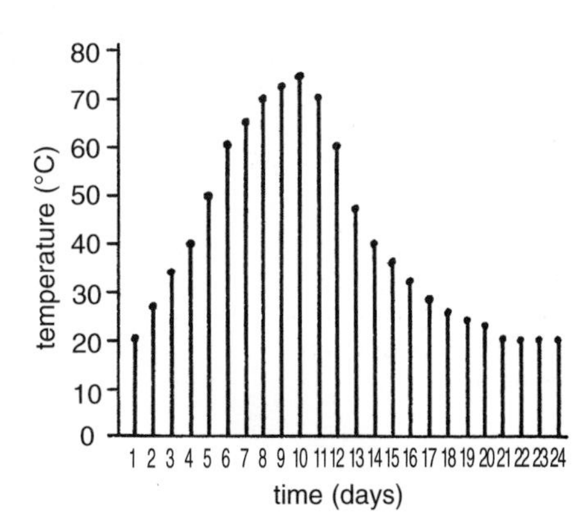

Items 9 and 10 refer to the following information. During a soil ecosystem investigation, certain members of the soil community were placed in containers like the one shown in the accompanying diagram. The results are given in the following table.

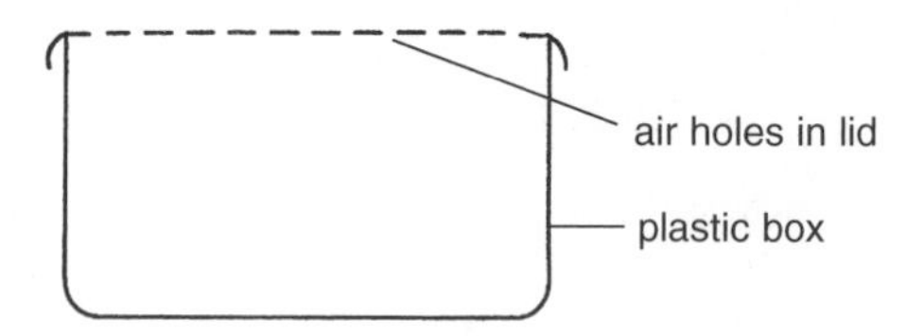

	Box 1	Box 2	Box 3	Box 4	Box 5	Box 6
Organisms present at start	2 lettuce roots and 2 centipedes	2 lettuce roots and 2 leather-jackets	2 lettuce roots and 2 millipedes	2 ground beetles and 2 leather-jackets	2 millipedes and 2 centipedes	2 lettuce roots and 2 ground beetles
Organisms present after 3 days	2 lettuce roots and 2 centipedes	2 leather-jackets	2 millipedes	2 ground beetles	2 centipedes	2 lettuce roots and 2 ground beetles

9 Which of the following conclusions is NOT justified from these results?

A Lettuce plants are producers and centipedes are secondary consumers.
B Leather-jackets and millipedes are both primary consumers.
C Millipedes and ground beetles are both preyed upon by centipedes.
D Ground beetles are predators and leather-jackets are their prey.

10 The reliability of the results could be increased by

A using a larger number of each type of soil organism.
B having only one primary consumer included in the experiment.
C omitting the ground beetles from the investigation.
D increasing the number of air holes in the lids of the boxes.

11 The accompanying diagram shows part of a seashore food web.

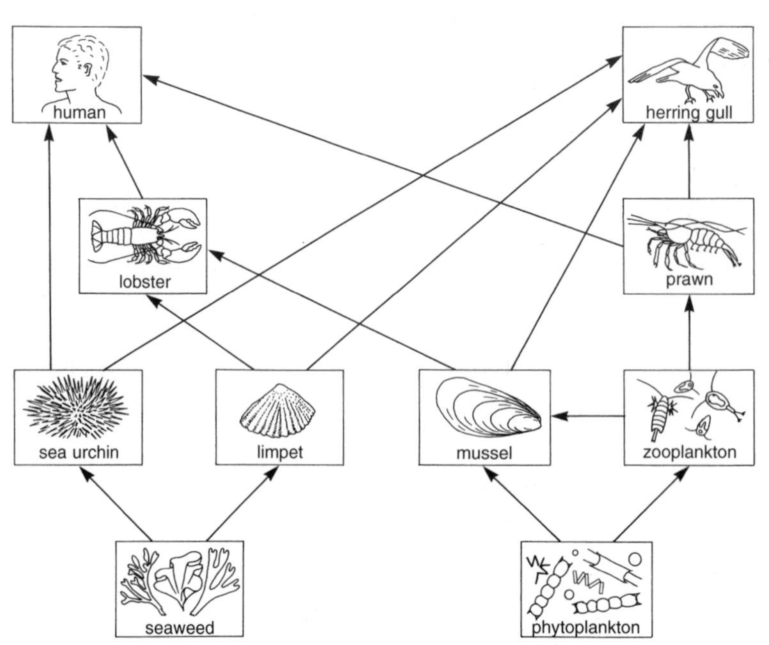

Which of the following food chains extracted from the seashore food web is INCORRECT?

A seaweed → limpet → lobster → human
B phytoplankton → mussel → lobster → human
C seaweed → sea urchin → lobster → herring gull
D phytoplankton → zooplankton → prawn → herring gull

Items 12, 13, 14 and 15 refer to the accompanying diagram which shows part of a food web in a freshwater pond.

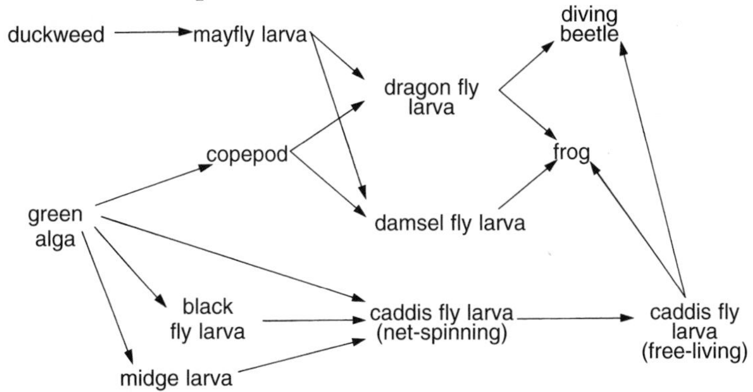

12 The number of different food chains that include dragonfly larva is

A 1. **B** 2. **C** 3. **D** 4.

13 An animal in this food web that is NOT eaten by other animals is

A diving beetle. **B** copepod. **C** midge larva. **D** mayfly larva.

14 An omnivore in this food web is

A green alga.
B blackfly larva.
C caddis fly larva (free-living).
D caddis fly larva (net-spinning).

15 If all the mayfly larvae were removed, it is very likely that the

A duckweed population would decrease.
B number of copepods would decrease.
C number of damsel fly larvae would increase.
D number of dragon fly larvae would increase.

16 The accompanying diagram shows a moorland food web where the arrows represent energy flow. Which arrow is INCORRECT?

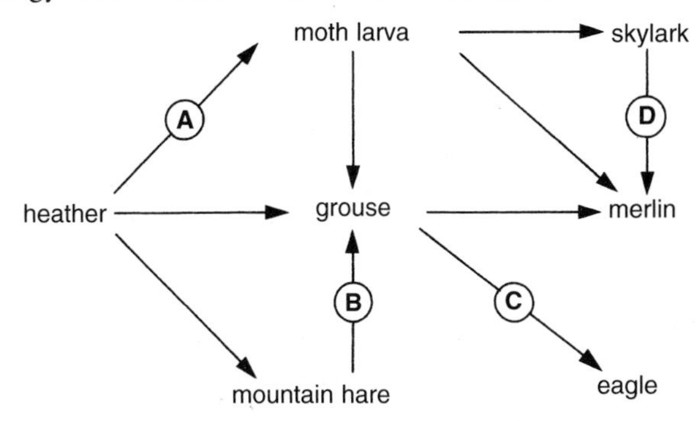

Items 17, 18 and 19 refer to the accompanying diagram which shows the number of units of energy in kJ/m² that were transferred from organism to organism in a grassland food chain.

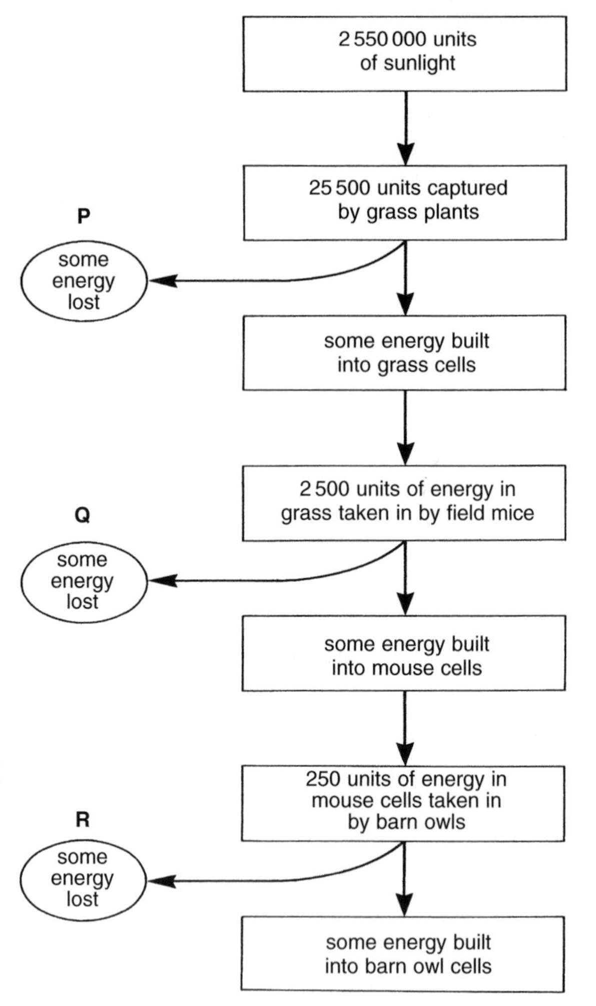

17 What percentage of sunlight energy was successfully captured by the grass?

A 1
B 10
C 90
D 99

18 What percentage of energy was lost between 'intake of energy in grass by field mice' and 'intake of energy in mouse cells by barn owls'?

A 1
B 10
C 90
D 99

19 Energy loss occurred at positions P, Q and R in the diagram. Energy lost in undigested food passed as faeces occurred at

A P and Q only.
B P and R only.
C Q and R only.
D P, Q and R.

Items 20 and 21 refer to the accompanying pie chart which shows the daily energy balance of a laying hen.

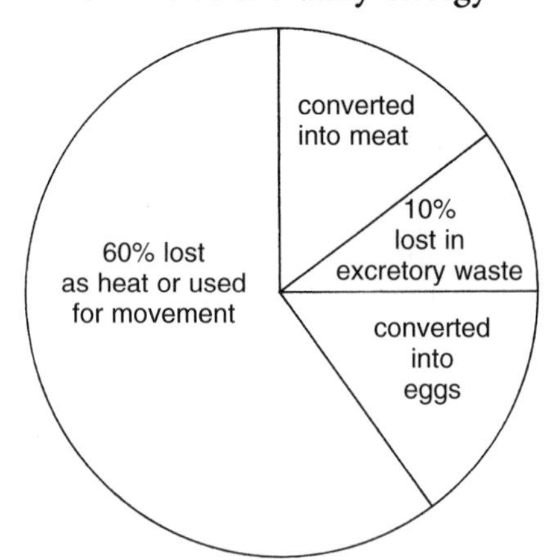

20 The percentage of the hen's energy intake that is converted into eggs is

A 15. **B** 20. **C** 25. **D** 30.

21 If the hen's daily total energy intake was 1220 kJ how many kilojoules were converted to meat?

A 122 **B** 183 **C** 305 **D** 366

22 An estimate of the biomass of each of the various types of organisms present in a freshwater ecosystem gave the results in the following table. Identify the producers.

Type of organism	Biomass (kg/hectare)
A	7
B	5192
C	56
D	623

23 The accompanying diagram shows part of a North Sea food web.

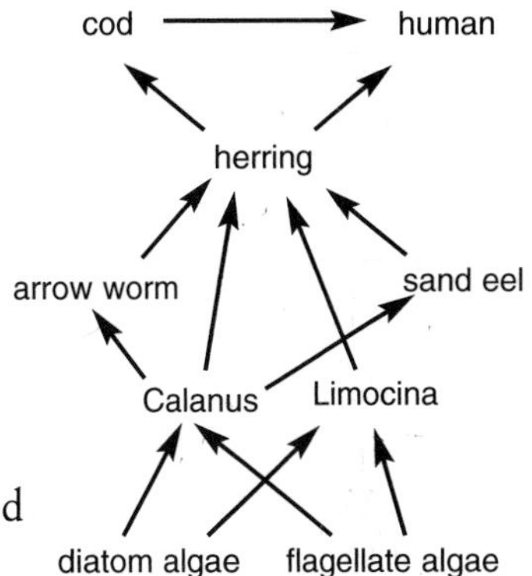

The second diagram shows four pyramids of biomass extracted from this food web. Which pyramid is INCORRECT?

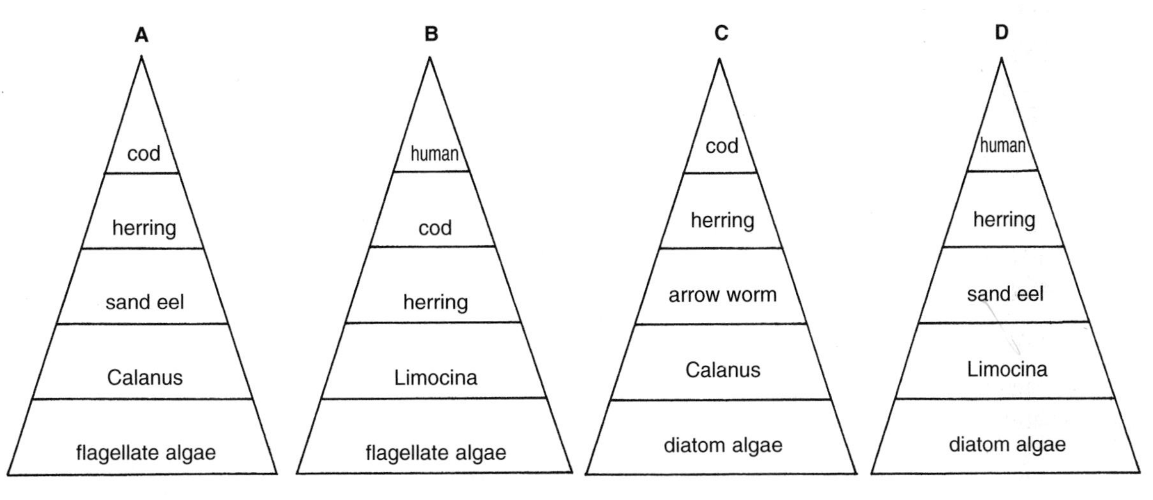

Items 24 and 25 refer to the following food chain and the four accompanying diagrams which comprise a set of possible answers.

grass → gazelle → lion → flea

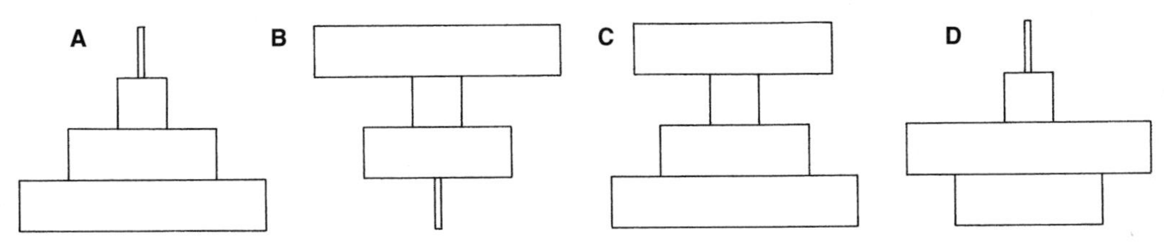

24 Which diagram best represents the food chain as a pyramid of numbers?

25 Which diagram best represents the food chain as a pyramid of energy?

Test 2

1 Within an ecosystem, the community consists of all the

A plant species only.
B animal species only.
C plant and animal species only.
D plant, animal and micro-organism species.

Items 2, 3 and 4 refer to the accompanying diagram which shows the flow of energy through a food chain.

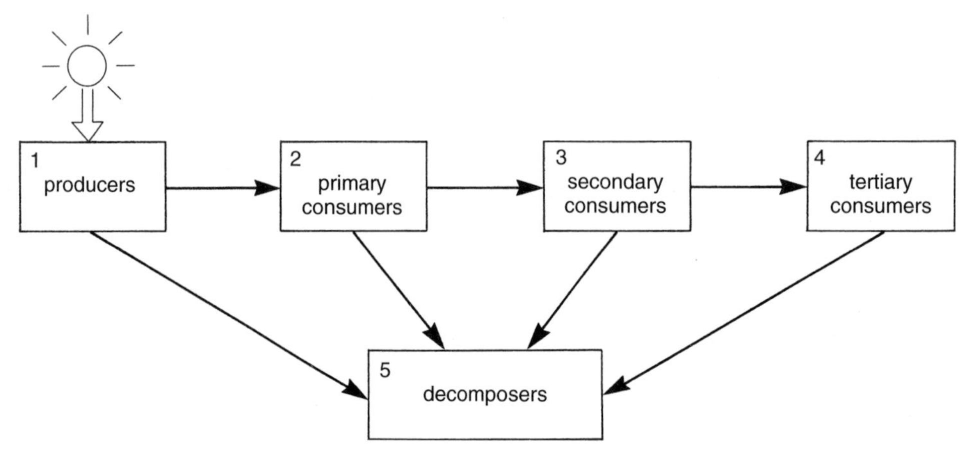

2 Which box refers to herbivores?

A 1 **B** 2 **C** 3 **D** 4

3 Which box refers to carnivores?

A 1 **B** 2 **C** 4 **D** 5

4 Which box refers to organisms that release chemical nutrients from waste materials?

A 1 **B** 3 **C** 4 **D** 5

5 The accompanying diagram shows a food web.

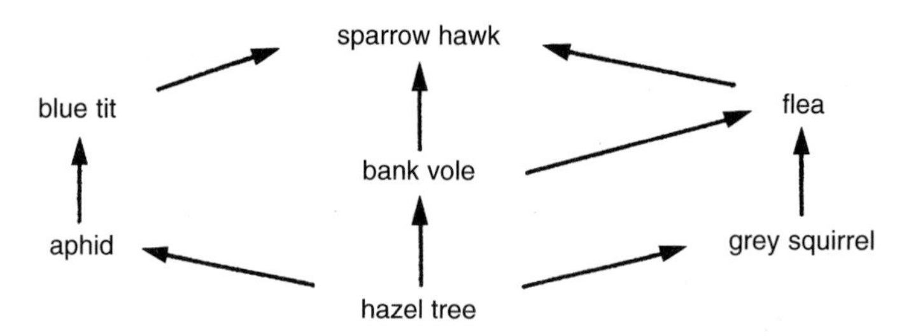

An example of a primary consumer in this food web is

A flea. **B** bank vole.
C hazel tree. **D** sparrow hawk.

Questions 6 and 7 refer to the accompanying diagram which shows a graph of the results from a study of a predator and its well-fed prey.

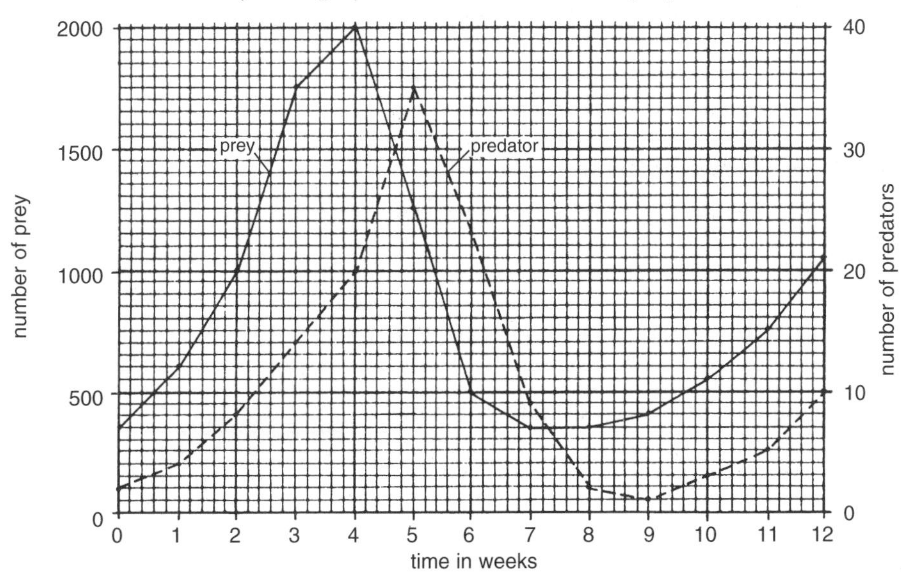

6 By how many individuals did the predator population increase in number from week 1 to week 5?

A 31 **B** 33 **C** 1550 **D** 1650

7 By how many times did the prey outnumber the predators at week 4?

A 2 **B** 25 **C** 100 **D** 1000

8 Six dishes were set up to investigate a soil food web as shown in the accompanying diagram.

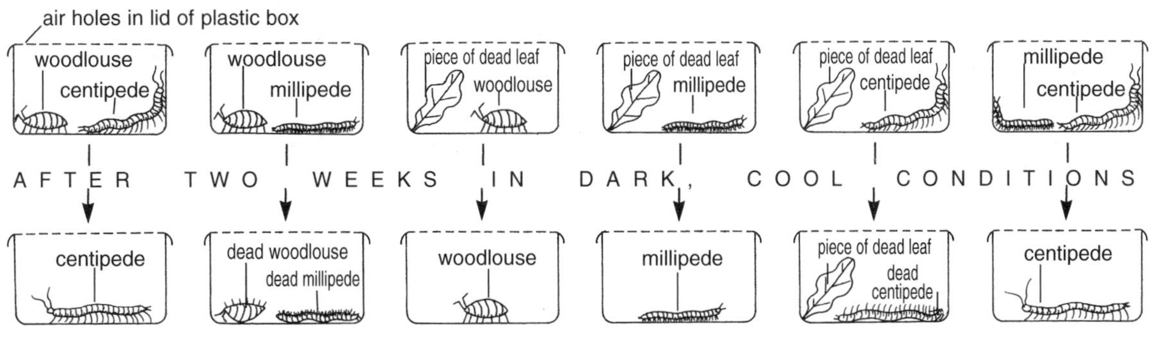

Which of the following food webs represents the feeding relationships that exist amongst these soil organisms?

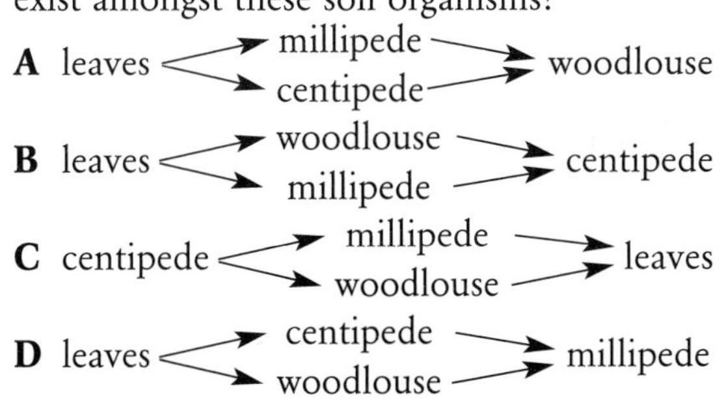

Questions 9, 10, 11 and 12 refer to the accompanying diagram of part of a marine food web.

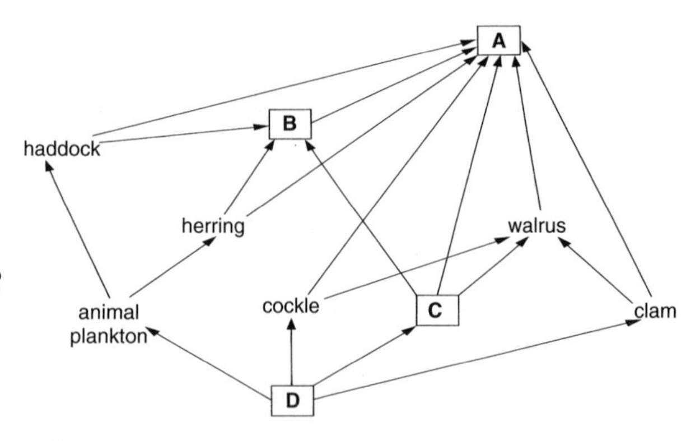

9 Which box should read 'mussel'?

10 Which box should read 'cod'?

11 Which box should read 'plant plankton'?

12 Which box should read 'human'?

Items 13, 14 and 15 refer to the accompanying diagram of part of a food web in a loch.

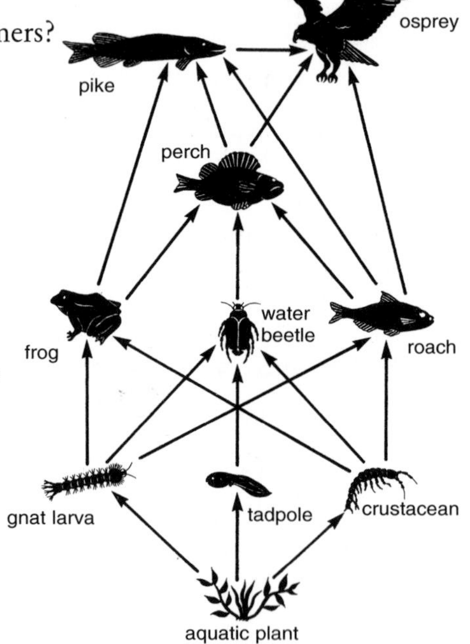

13 Which of the following are BOTH secondary consumers?

A perch and frog
B roach and perch
C frog and water beetle
D water beetle and tadpole

14 The number of different primary consumers in this food web is

A 1. **B** 3. **C** 6. **D** 9.

15 The population of organisms with the largest relative biomass would be

A pike.
B perch.
C osprey.
D aquatic plants.

Questions 16 and 17 refer to the prehistoric food web shown in the accompanying diagram.

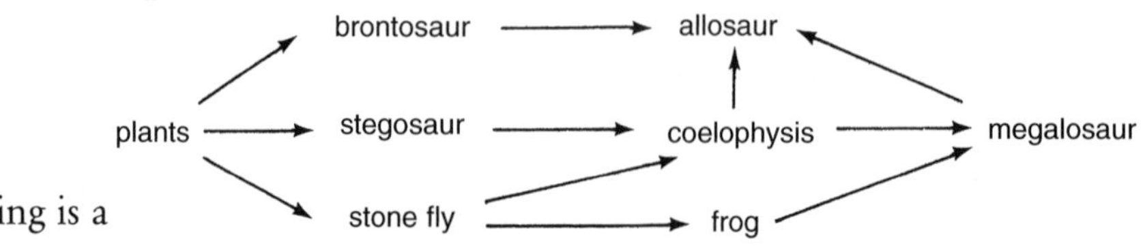

16 Which of the following is a tertiary consumer?

A allosaur **B** stegosaur
C coelophysis **D** frog

17 If brontosaur became extinct, which of the following would occur first?

A an increase in the number of coelophysis
B an increase in the number of stegosaur
C a decrease in the number of frogs
D a decrease in the number of stone flies

18 The energy content for the members of a grassland community was calculated as shown in the following table.

Type of organism	Energy equivalent (kJ/m²/year)
producers	5500.0
primary consumers	330.1
secondary consumers	40.0
tertiary consumers	3.8

9.5% of available energy was transferred from

A sunlight to producers.
B producers to primary consumers.
C primary consumers to secondary consumers.
D secondary consumers to tertiary consumers.

19 A loch was stocked with 200 one-year-old trout each weighing, on average, 30 g. During the following year the trout increased in mass by 50% and then 40% of the trout were caught by fishermen. The total biomass in grams of trout remaining in the loch was

A 1200. **B** 1800. **C** 3600. **D** 5400.

Items 20 and 21 refer to the accompanying diagram which shows the fate of the energy present in poultry feed consumed by a laying hen.

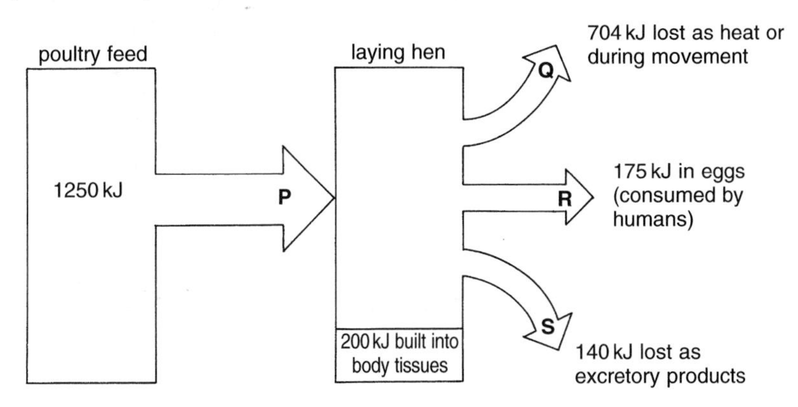

20 The percentage of energy in poultry feed successfully converted to the hen's body tissues is

A 1.60. **B** 6.25. **C** 16.00. **D** 626.00.

21 Which arrow represents a source of energy readily available to decomposers?

A P **B** Q **C** R **D** S

22 Which of the following diagrams correctly represents a pyramid of numbers?

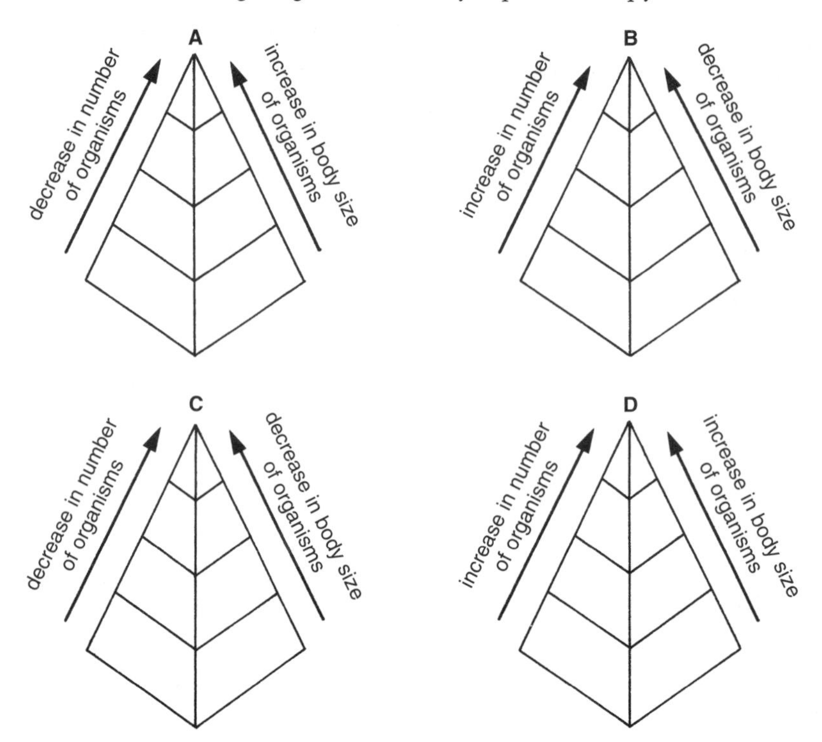

23 The following list gives five food chains. If the biomass of the producer is equal in each, which TWO could support the largest number of human beings?

1 wheat → human
2 corn → cow → human
3 rice → human
4 corn → hen → human
5 wheat → pig → human

A 1 and 3 **B** 1 and 5
C 2 and 3 **D** 3 and 4

24 The accompanying diagram shows a pyramid of biomass for an area of grassland.

The information missing from parts X and Y should read

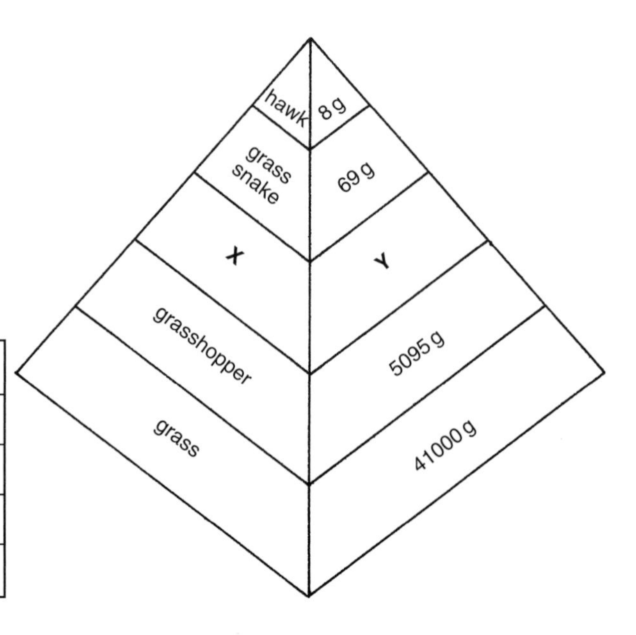

	X	Y
A	frog	570 g
B	magpie	570 g
C	frog	5705 g
D	magpie	5705 g

25 The accompanying diagram shows a woodland food web.

The second diagram shows four pyramids of **numbers** extracted from this food web. Which one is CORRECT?

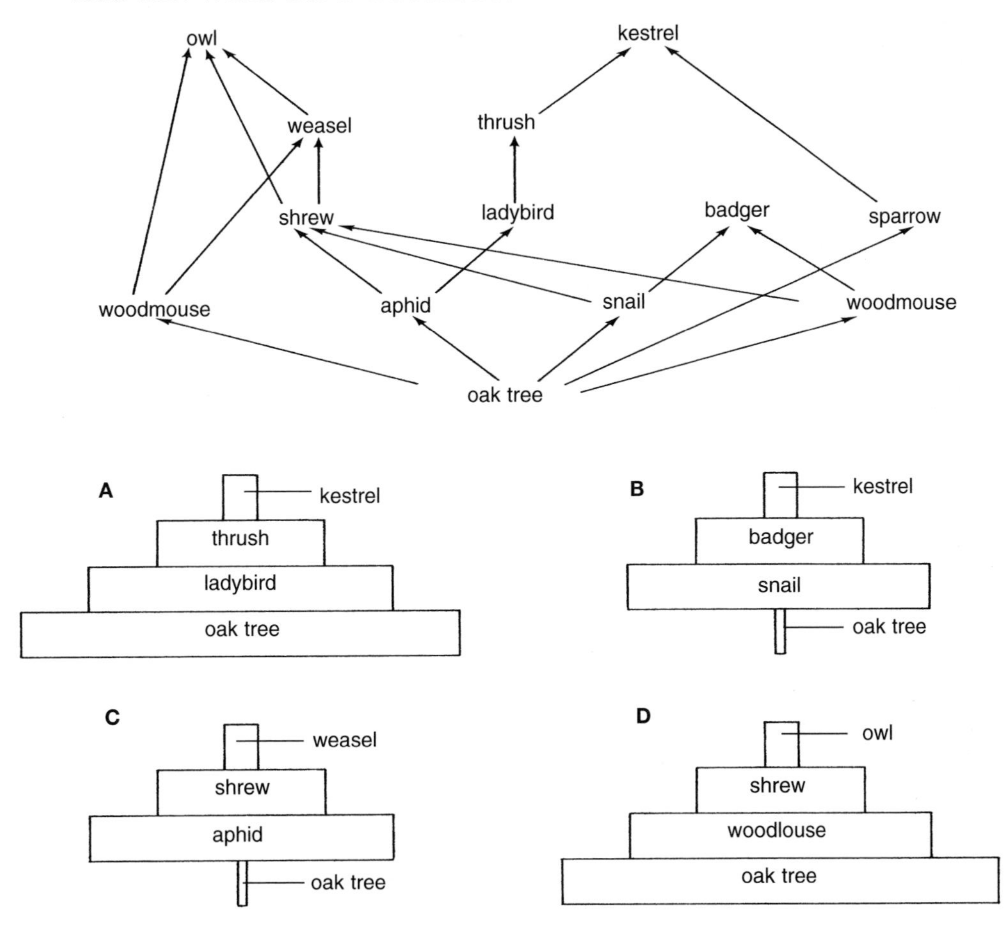

Factors affecting the variety of species in an ecosystem

Test 1

1 Which of the following statements is FALSE?

Members of the same species

A compete with one another when food is scarce.
B are able to interbreed and produce fertile offspring.
C are always found living together in the same ecosystem.
D all possess the same set of chromosomes in their cells.

2 Which of the following terms refers to an organism's whole way of life and the use to which it puts the available environmental resources?

A niche
B habitat
C community
D ecosystem

3 Biodiversity is best defined as the variation

A that exists amongst the members of both sexes within a single species.
B produced amongst the offspring of a cross between two members of a species.
C that occurs amongst members of the same species adapted to different ecosystems.
D found within and between all species present in the world's many different ecosystems.

4 Which of the following is a behavioural adaptation?

A the strong, hard shell of a barnacle that resists damage by wave action
B the decrease in rate of activity of a blowfly larva that it shows in dark conditions
C the yellow pigments in a shade plant's leaves that help it to photosynthesise in dim light
D the strong holdfast of a brown seaweed that attaches it firmly to submerged rocks

5 Which of the maps in the accompanying diagram shows the most likely worldwide distribution of mosquitoes following a global warming of 5 °C in the future? (Note: an isotherm is a contour line of equal temperature.)

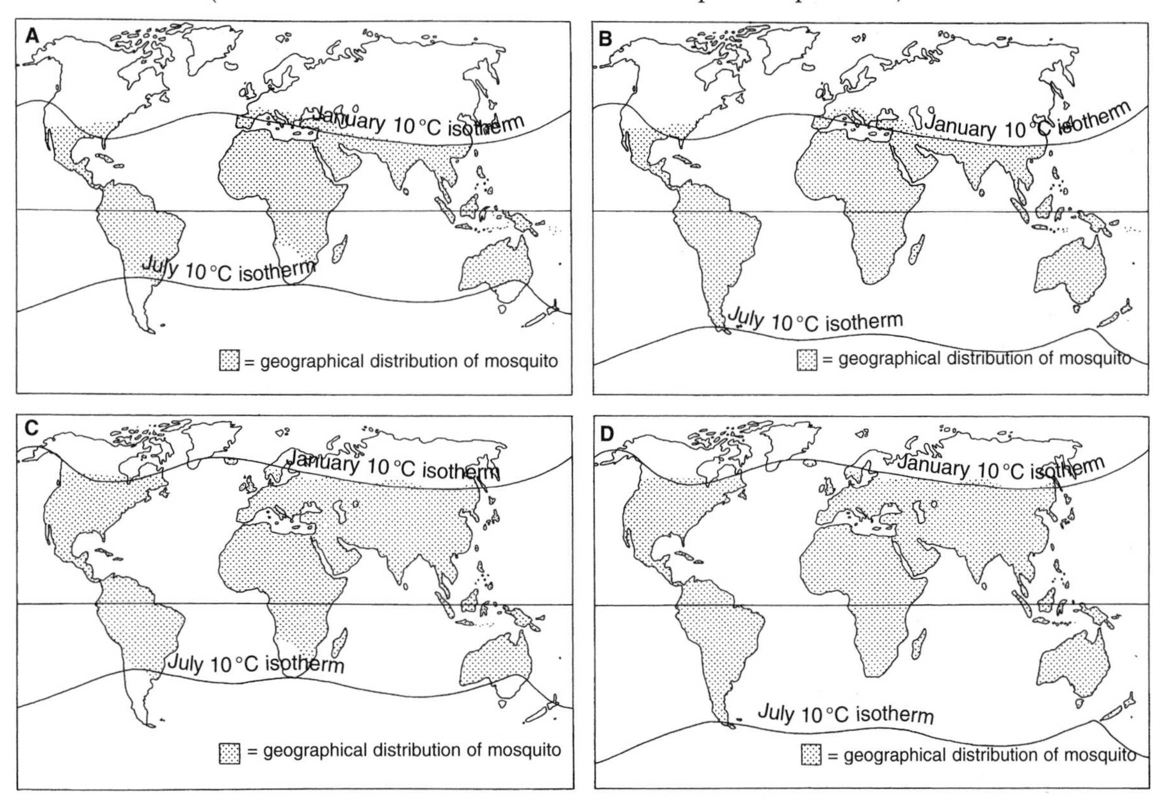

Items 6 and 7 refer to the accompanying kite diagram which shows the effect of increasing depth on five environmental factors typical of a freshwater loch.

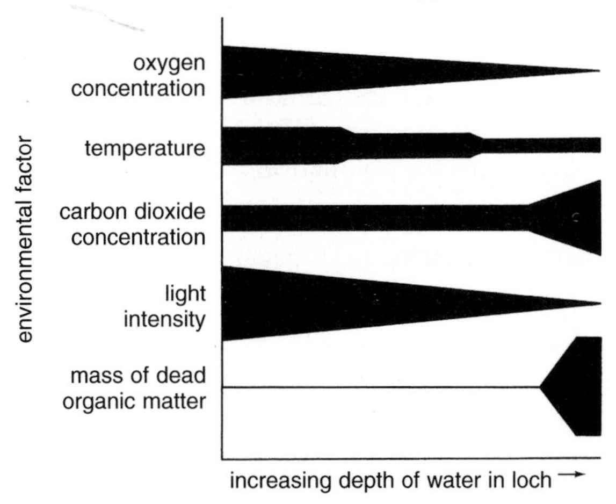

6 Algae are found growing close to the surface of the loch. The factor that restricts the distribution of the algae to this region of the water is the

A concentration of oxygen.
B concentration of carbon dioxide.
C intensity of light.
D mass of dead organic matter.

7 Most of the loch's population of bacteria are found at the bottom. The factor mainly responsible for this distribution of the bacteria is the

A concentration of oxygen.
B temperature.
C concentration of carbon dioxide.
D mass of dead organic matter.

8 The accompanying diagram shows a rocky shore. An abundance survey of the periwinkle populations was carried out at sites W, X, Y and Z. The results are summarised in the table where 0 = absent, 1 = rare, 2 = occasional, 3 = frequent and 4 = abundant.

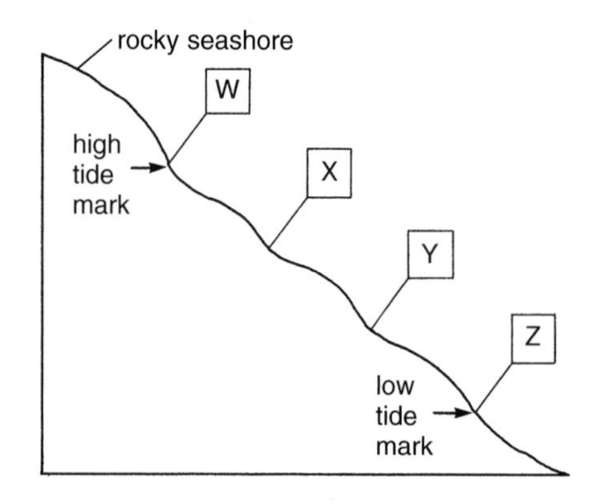

	Location			
	W	**X**	**Y**	**Z**
rough periwinkle	2	3	1	0
small periwinkle	4	2	0	0
flat periwinkle	0	1	4	2

From this data it can be concluded that

A rough periwinkles are most numerous near the low tide mark.
B small periwinkles are most numerous near the low tide mark.
C flat periwinkles are adapted to survive long periods out of the water.
D small periwinkles are adapted to survive long periods out of the water.

9 In a stable ecosystem, a wide variety of

A animals depend on plants for food and oxygen.
B producers depend on plants for shelter and camouflage.
C micro-organisms depend on plants for carbon dioxide and nitrogen.
D plants depend on micro-organisms for pollination and seed dispersal.

10 The following list gives four situations that could affect a mixed community of plants growing in a grassland ecosystem.

1 Herbivores graze unselectively on all species.
2 All herbivores are removed from the ecosystem.
3 Herbivores graze selectively on delicate plant varieties.
4 Herbivores graze selectively on sturdy dominant grasses.

Which TWO situations would tend to maintain species diversity amongst the plant community?

A 1 and 3 **B** 2 and 3
C 1 and 4 **D** 2 and 4

Items 11 and 12 refer to the accompanying chart which shows the effect of pH on the biodiversity of fish species in a freshwater loch.

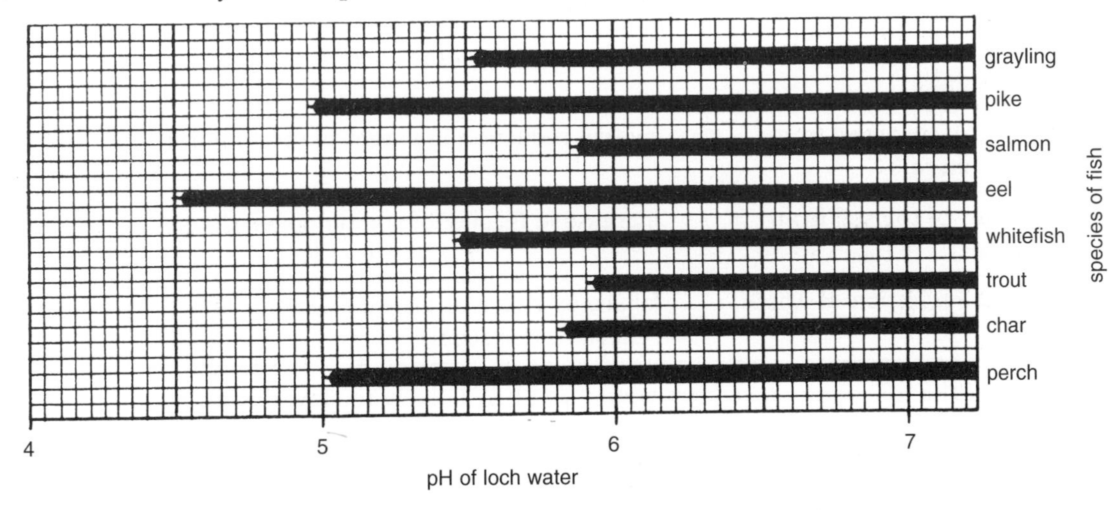

11 The number of species of fish that are found in loch water of pH 5.25 is

A 1. **B** 3. **C** 4. **D** 5.

12 No salmon are found at pH

A 5.75. **B** 5.95. **C** 6.05. **D** 6.25.

Items 13 and 14 refer to the accompanying graph which charts the oxygen concentration of the water in a river from where it meets the sea back to its source 44 miles inland. This river has suffered pollution by untreated sewage.

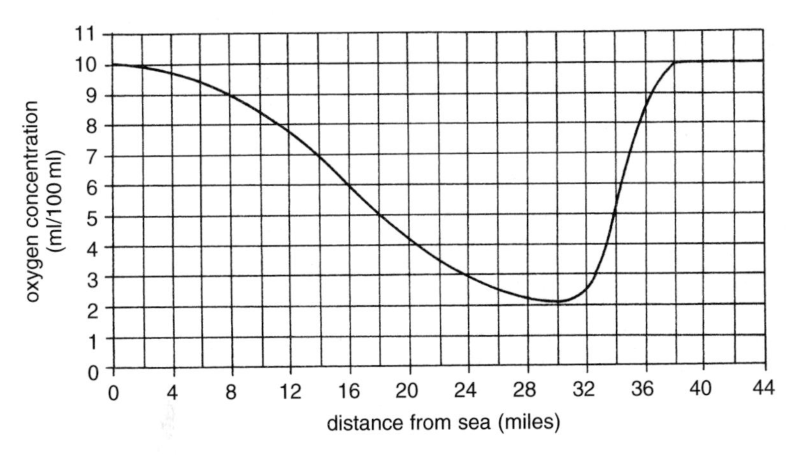

13 The distance from the sea, in miles, at which the organic effluent was added was

A 2. **B** 30. **C** 38. **D** 44.

14 Stonefly nymphs are adapted to life in oxygen concentrations of more than 3 ml per 100 ml of water. The distance from the sea, in miles, at which stonefly nymphs would be found is

A 22. **B** 26. **C** 28. **D** 32.

15 Which line in the following table is CORRECT?

	Effect of deforestation	Possible consequences
A	less water vapour returned to atmosphere	increase in greenhouse effect
B	less CO_2 absorbed from atmosphere	overall reduction in extent of local rainfall
C	erosion of top soil from hillsides	increase in fertility of soil on upper slopes
D	lack of regulation of water flow	failure by rivers to provide humans with reliable water supply

16 Which of the following procedures would be LEAST likely to slow down the process of desertification?

A planting trees to act as a windbreak
B cultivating marginal land for crop-growing
C conserving resilient grasses that prevent soil erosion
D pursuing traditional farming practices such as crop rotation

17 At the end of the twentieth century all of the following animals were in danger of becoming extinct EXCEPT

A red deer. **B** blue whale.
C grizzly bear. **D** mountain gorilla.

18 Grey seals were hunted so intensively before the *Grey Seal Protection Act* of 1914 that only about 1800 individuals remained in Scotland at that time.

By 1999, Scotland possessed about 45% of the world's total grey seal population of 200 000. During the 20th Century the Scottish population of grey seals had increased by a factor of

A 45. **B** 50. **C** 90. **D** 111.

Items 19 and 20 refer to the accompanying bar graphs which compare the number of species of vertebrate animals and vascular plants recorded in the years 1900 and 2000 for a large region of mainland Scotland.

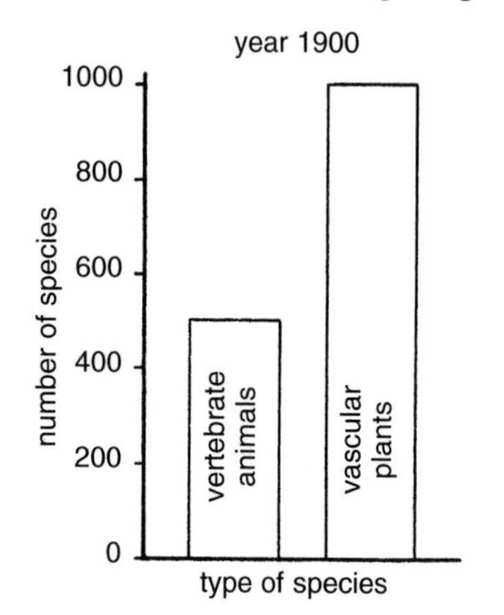

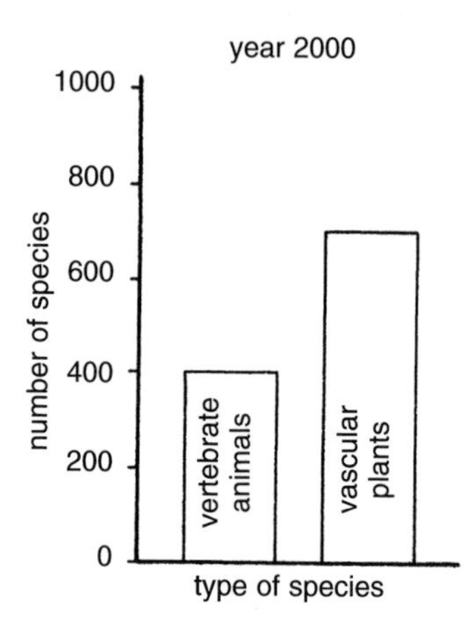

19 The percentage decrease in species number of each group is

	Vertebrate animals	Vascular plants
A	10	30
B	20	30
C	10	70
D	20	70

20 The factor most likely to have been responsible for these changes in biodiversity is

A global warming.
B gale-force winds.
C agricultural intensification.
D glaciation during an ice age.

21 Which of the following is an aesthetic reason for conserving biodiversity?

A Many species are beautiful and enrich our lives.
B Mixed hedges on farmland often act as windbreaks.
C Some grasses on sand dunes reduce coastal erosion.
D Tough species of seaweed reduce wave action on the coastline.

22 Which line in the table is CORRECT?

	Animal	Environmental stimulus	Behavioural response and adaptive significance
A	rabbit	increasing daylengths	animals mate in autumn to ensure that young are born in spring
B	swallow	increasing daylengths	animals hibernate to avoid cold winter conditions
C	hedgehog	decreasing daylengths	animals migrate to survive unfavourable winter conditions
D	fiddler crab	rhythmical tidal movement	animals emerge at low tide to look for food

23 A choice chamber was set up so that the air in one side was damp and the air in the other side was dry. Twenty woodlice were then released into the choice chamber. The results were graphed as shown in the accompanying diagram.

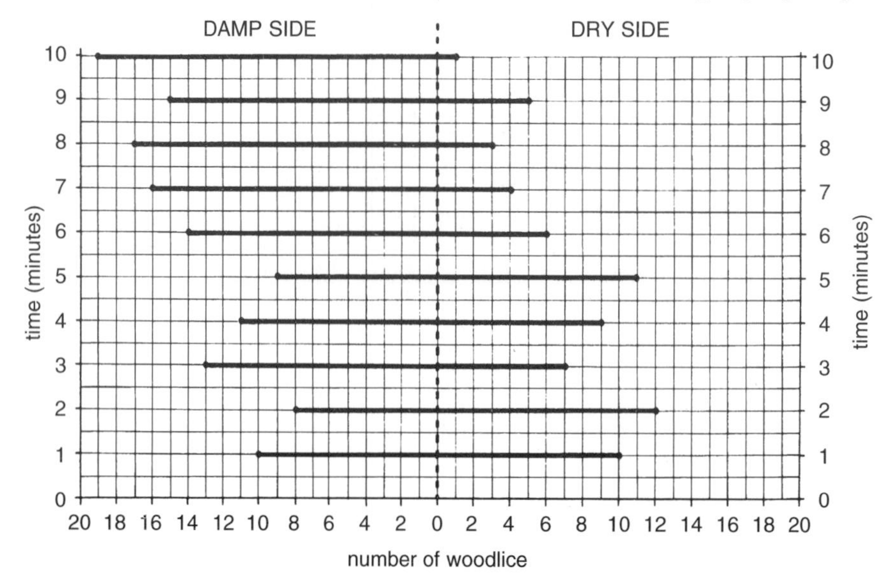

Which of the following statements is CORRECT?

A At minute 3, 13 woodlice were in the dry side.
B At minute 5, 55% of the woodlice were in the damp side.
C At minute 7, 1 in 5 of the woodlice were in the dry side.
D At minute 9, the ratio of woodlice in the damp to those in the dry side was 4:1.

24 *Galium saxatile* grows best in acidic soil but will grow on alkaline soil in the absence of competition. *Galium pumilum*, on the other hand, grows best on alkaline soil but can survive on acidic soil.

In a competition experiment, equal numbers of seeds of both species of *Galium* were planted together in two pots. Pot 1 contained alkaline soil and pot 2 contained acidic soil.

Which of the following sets of results is the most likely outcome of this experiment?

	Pot I (alkaline soil)		Pot 2 (acidic soil)	
	G. saxatile	***G. pumilum***	***G. saxatile***	***G. pumilum***
A	+	-	-	+
B	-	+	+	-
C	+	-	+	-
D	-	+	-	+

+ = growth - = no growth

25 The accompanying graph shows the results of a competition experiment between two species of animal (X and Y) that consume the same type of food.

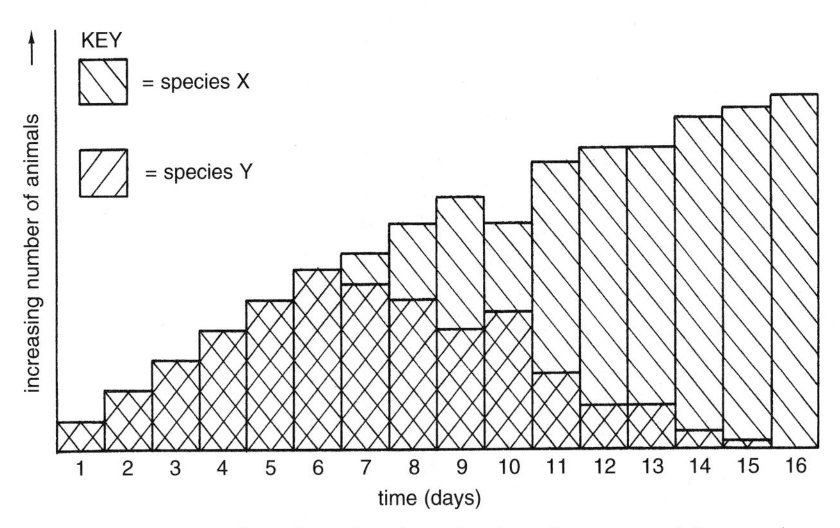

Species X was first found to be winning the competition at day

A 6. **B** 7. **C** 12. **D** 16.

Test 2

Items 1 and 2 refer to the following table.

	Habitat	Niche
A	woodland	omnivorous consumer suffering intense competition from introduced rival
B	seashore	producer that provides food and shelter for many marine animals
C	woodland	producer adapted to life in dimly lit conditions
D	seashore	herbivorous consumer preyed upon by predators such as crabs

1 Which line in the table refers to limpet?

2 Which line in the table refers to bluebell?

3 Which structure in the accompanying diagram is an adaptation which helps Scots pine to survive in windy environments by reducing water loss?

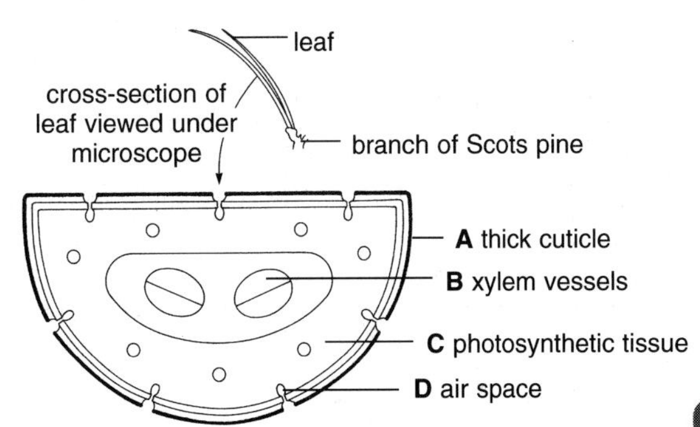

Items 4, 5 and 6 refer to the accompanying diagram which charts the results from an experiment to investigate the effect of soil pH on the distribution of four plant species (W, X, Y and Z).

The soil pH and percentage cover of each plant species were recorded at each of ten sample points at regular intervals along a line transect. The plant cover data are presented as kite diagrams.

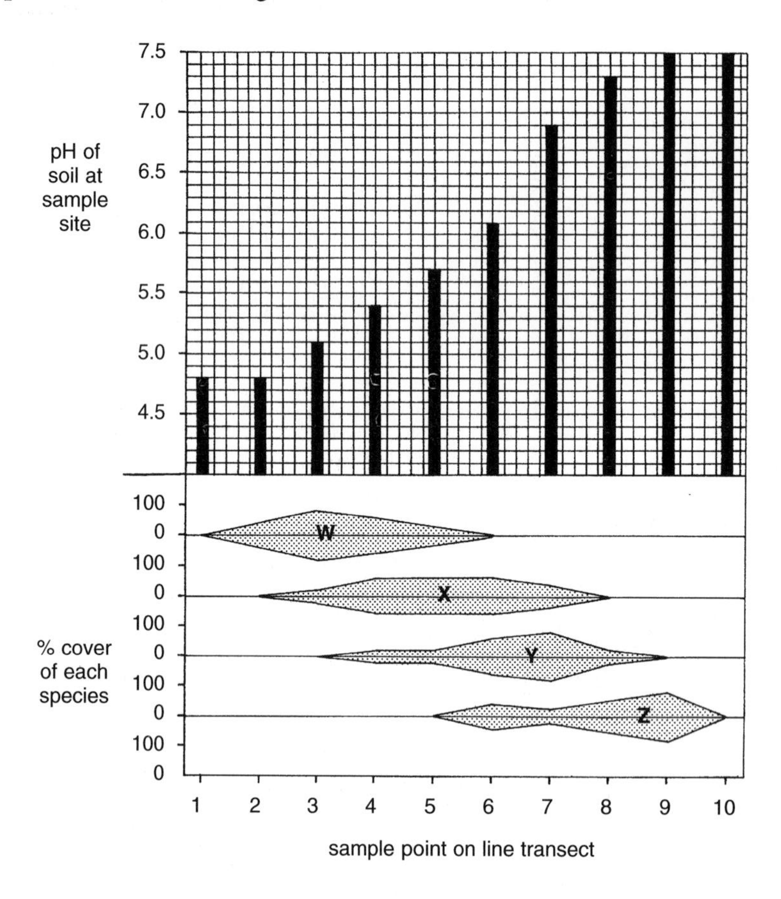

4 Only two of the plant species were recorded at sample point

A 3. **B** 4. **C** 5. **D** 6.

5 Species W, X and Y were all found to be growing in soil of pH

A 5.1 **B** 5.7 **C** 6.1 **D** 6.9

6 Which species is able to tolerate the widest range of soil pH?

A W **B** X **C** Y **D** Z

7 The accompanying diagram represents a stable ecosystem and the interdependent members of its community. Which of the following is flowing through the ecosystem at ALL of arrows 1, 2, 3 and 4?

A energy
B oxygen
C mineral salts
D carbon dioxide

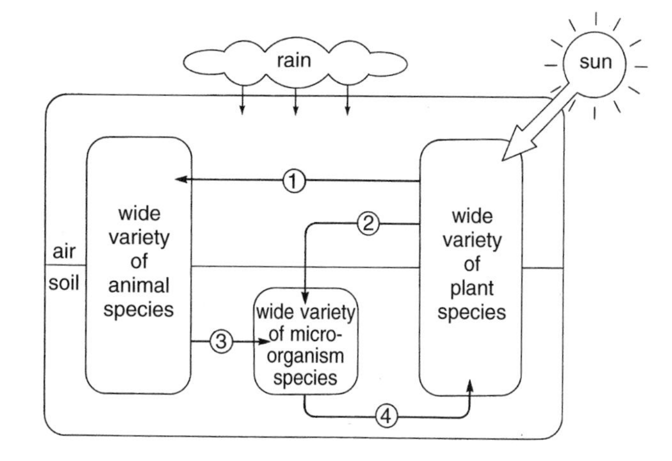

Items 8, 9 and 10 refer to the accompanying kite diagram. It charts the results from a survey of the range and abundance of seven plants and animals present on a rocky seashore.

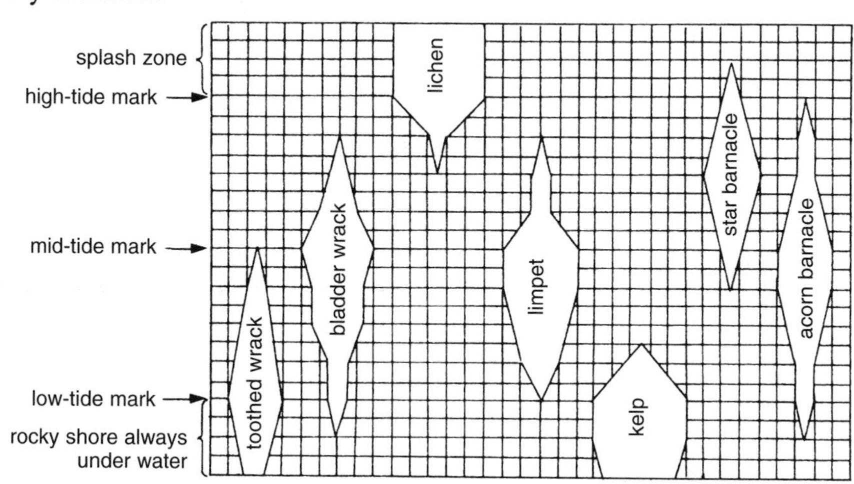

8 Which line in the following table is CORRECT?

	Location of greatest abundance	
	Toothed wrack	**Bladder wrack**
A	low-tide mark	mid-tide mark
B	mid-tide mark	low-tide mark
C	mid-tide mark	mid-tide mark
D	low-tide mark	low-tide-mark

9 From the data, the organism LEAST able to tolerate long periods out of water appears to be

A kelp. **B** lichen.
C toothed wrack. **D** star barnacle.

10 Which organism is able to survive the widest range of environmental conditions on this rocky seashore?

A kelp **B** limpet
C bladder wrack **D** acorn barnacle

Items 11 and 12 refer to accompanying graph which shows the results from a survey done on the number of lichen species growing along a 20 km transect, from the centre of a city out to a country area.

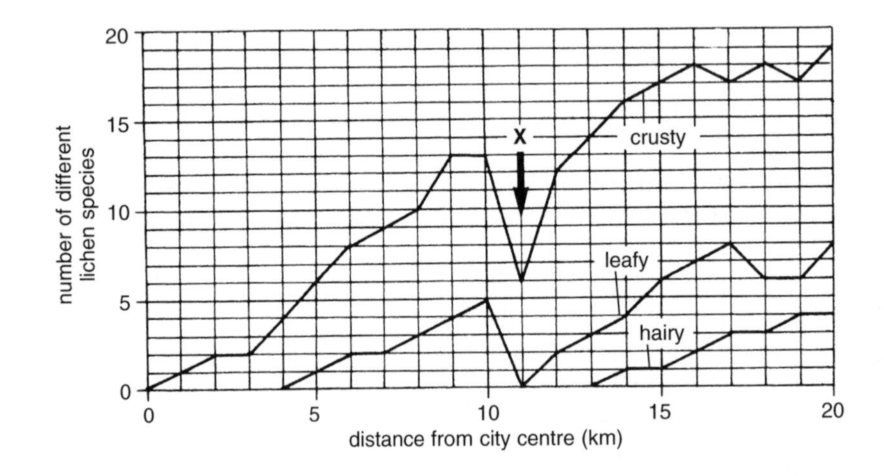

11 The dip in the graph at arrow X indicates

A an area of especially clean air.
B an area lacking both hairy and crusty lichens.
C a local increase in sulphur dioxide concentration of air.
D a lower level of atmospheric pollution compared with the country area.

12 Twenty-eight different species of lichen were recorded at one of the sites. The distance (in km) of this site from the city centre was

A 16. **B** 17. **C** 18. **D** 19.

13 The following table shows the effect of pollution on the biodiversity of a river ecosystem.

	——— Direction of flow of river water ———→			
	State of water			
	Clean	**Very badly polluted**	**Partly polluted**	**Clean**
Biodiversity present in ecosystem	large number of different species	very small number of different species	**X**	large number of different species

↑ untreated sewage added (between Clean and Very badly polluted)

Which of the following should have been inserted in box **X** in the table?

A very small number of different species
B small number of different species
C large number of different species
D very large number of different species

14 Some of the events that lead to desertification are listed below.

1 Forests are cleared and pastures are overgrazed.
2 Agricultural land is lost to advancing sand dunes.
3 The human population increases rapidly.
4 Topsoil dries out and is blown away.

The sequence in which these events occur is

A 1, 3, 2, 4. **B** 1, 3, 4, 2.
C 3, 1, 2, 4. **D** 3, 1, 4, 2.

15 All of the following tend to reduce the biodiversity present in an ecosystem EXCEPT

A habitat conservation.
B atmospheric pollution.
C disruption of simple food webs.
D selective grazing of less sturdy species.

16 Which of the following animals was closest to worldwide extinction at the end of the 20th Century?

A lion **B** wolf
C zebra **D** panda

17 The accompanying diagram shows the effect of removing starfish (the dominant predator) from a marine food web.

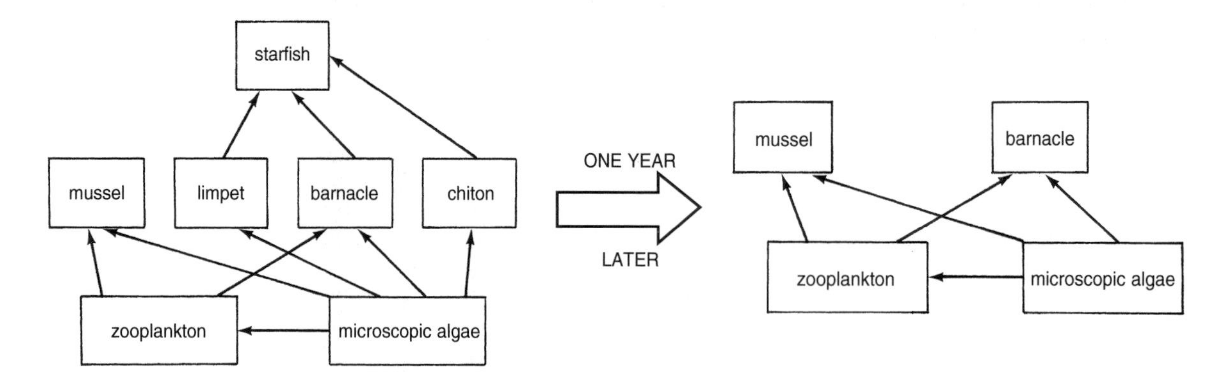

Which of the following statements is FALSE?

In the absence of starfish

A limpets and chitons lost out due to competition and died.
B barnacles remained uneaten and increased in number.
C limpets and chitons were deprived of zooplankton by barnacles.
D barnacles won out in the competition for algae as food.

18 Which of the following is an economic reason for conserving biodiversity?

A It is morally wrong to allow thousands of species to become extinct.
B The successful functioning of natural balanced ecosystems requires stability.
C It may be possible to develop new foods from wild varieties of plants.
D The variety of wild life species helps to enrich the nation's landscape.

19 Which of the following is an example of a desert animal exhibiting behavioural adaptation?

A the lizard which spends time in the shade to lower its body temperature
B the rattlesnake which possesses a scaly, waterproof body covering
C the kangaroo rat which has kidneys able to make very concentrated urine
D the camel which has long thick eye lashes for protection during sand storms

Items 20, 21 and 22 refer to the following information. Twenty woodlice were released into a choice chamber with one side illuminated and one side in darkness. The distribution of the woodlice was recorded each minute for 10 minutes. The results are shown in the accompanying bar chart.

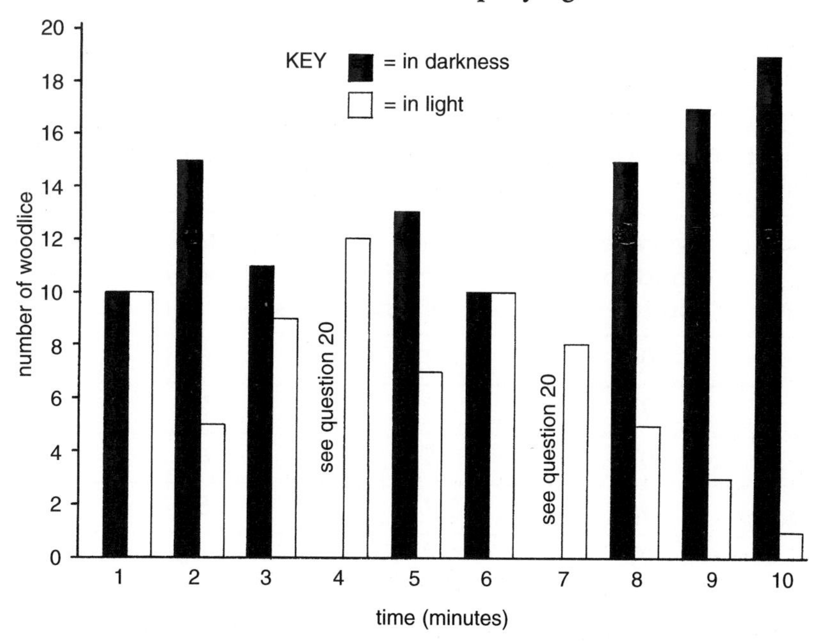

20 Which line in the accompanying table supplies the data needed to complete the bars missing from the chart at minutes 4 and 7?

	Number of woodlice in darkness at minute:	
	4	**7**
A	8	8
B	8	12
C	12	8
D	12	12

21 Expressed as a ratio, the number of woodlice in the dark to those in the light at minute 8 is

A 5:1. **B** 4:1. **C** 3:1. **D** 2:1.

22 The percentage of woodlice in darkness at minute 5 is

A 6.5. **B** 13.0. **C** 35.0. **D** 65.0.

23 The apparatus in the accompanying diagram was used to investigate the effect of light and dark on blowfly larvae. It shows the results 10 minutes after 4 larvae were inserted at point X.

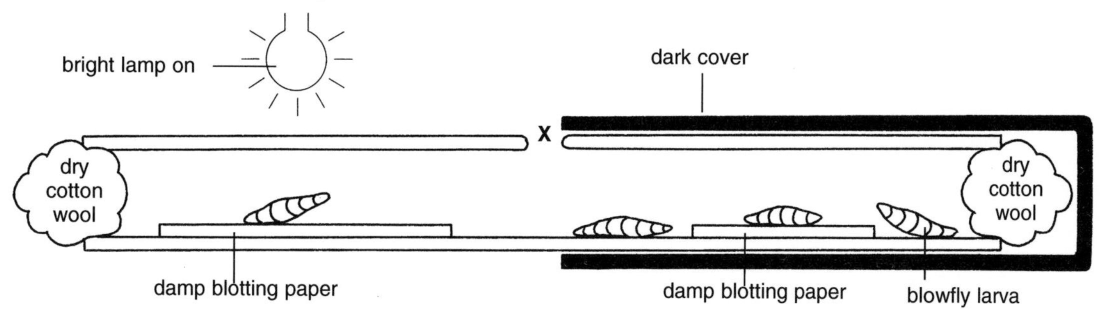

Which of the following is NOT a feature of the experimental design that needs to be altered to make the experiment valid?

A the location of the bright lamp
B the total number of blowfly larvae used in the investigation
C the mass of blotting paper used in each side of the apparatus
D the unequal number of animals in the two sides of the chamber

24 The table below gives the results from a plant competition experiment where 5 groups of pea plants were grown in areas of fertile soil measuring 0.25 m^2.

Plant group	Number of plants per 0.25 m^2	Average length of pod (mm)	Average number of pods per plant	Average number of seeds per pod	Average mass of seed (g)
1	20	100	8.3	6.0	0.66
2	40	98	6.8	5.9	0.54
3	60	102	3.9	6.2	0.69
4	80	95	2.7	5.9	0.75
5	100	97	2.1	6.0	0.63

From the data, which feature appears to be affected by competition between neighbouring pea plants?

A average length of pod
B average number of pods per plant
C average number of seeds per pod
D average mass of seed

25 The accompanying diagrams show the same area of moorland that contains territories inhabited by red grouse in different years. Which diagram BEST represents the year in which the food supply was most plentiful?

A

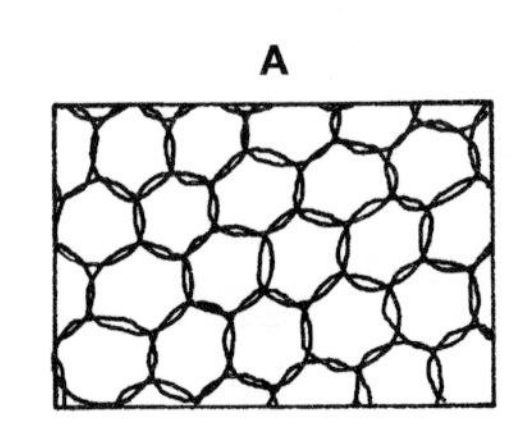

B

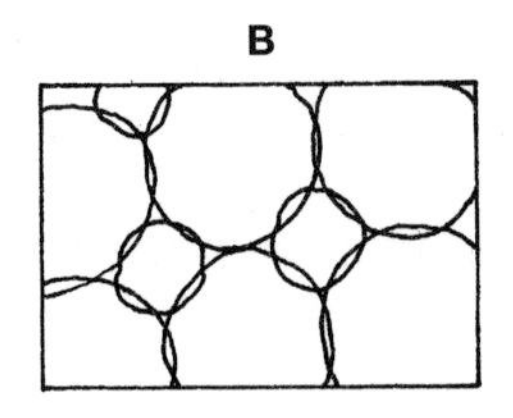

C

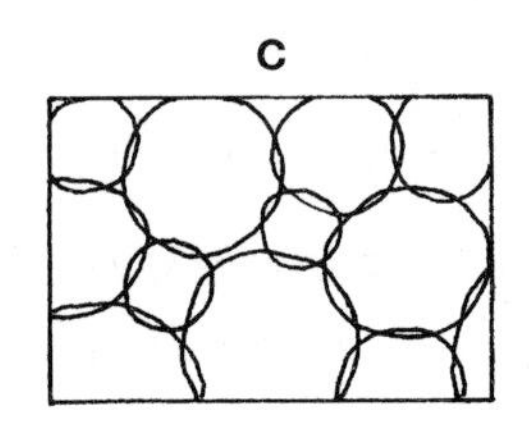

D

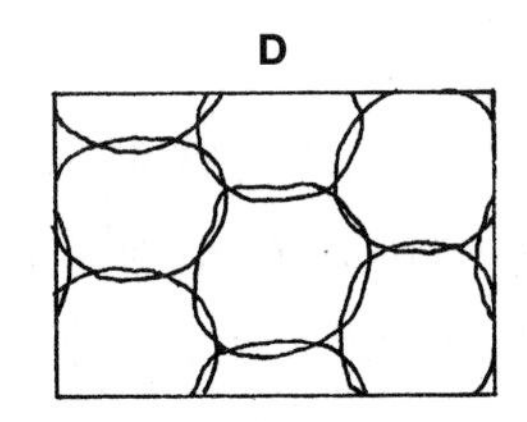

8 Factors affecting variety in a species

Test 1

1 Which line in the following table is CORRECT?

	Site of sperm production	Site of egg production
A	penis	oviduct
B	testes	oviduct
C	penis	ovary
D	testes	ovary

2 On average, the length of a human sperm is 0.06 mm. Expressed in micrometres, this would be

A 6. **B** 60. **C** 600. **D** 6000.

3 A man is considered to be fertile if his sperm count is 20 million or more sperm per millilitre (ml) of semen. The data in the accompanying table refer to four patients attending a fertilisation clinic.

Which patient is MOST likely to be infertile?

Patient	Average volume of semen (ml)	Average total number of sperm in semen (millions)
A	3.5	77
B	4.0	80
C	4.5	81
D	5.0	105

Questions 4, 5 and 6 refer to the accompanying diagram of part of a flower.

4 Which letter indicates the location of an egg cell?

5 Which letter shows the site of pollen grain production?

6 Which letter indicates the point where a zygote could be formed?

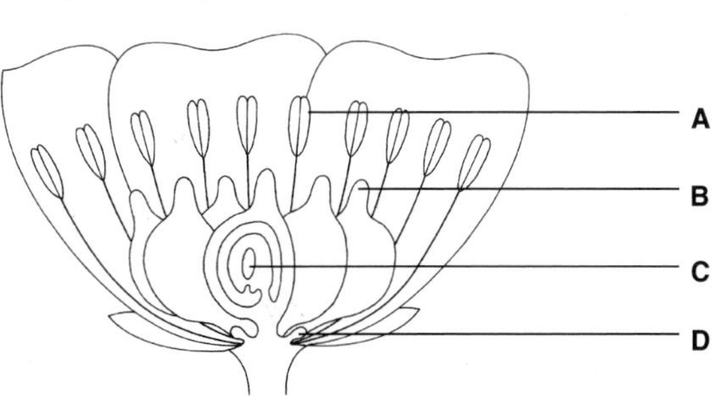

7 The first of the accompanying diagrams shows a small part of a DNA molecule where the four types of base molecule are represented by the letters A, T, G and C.

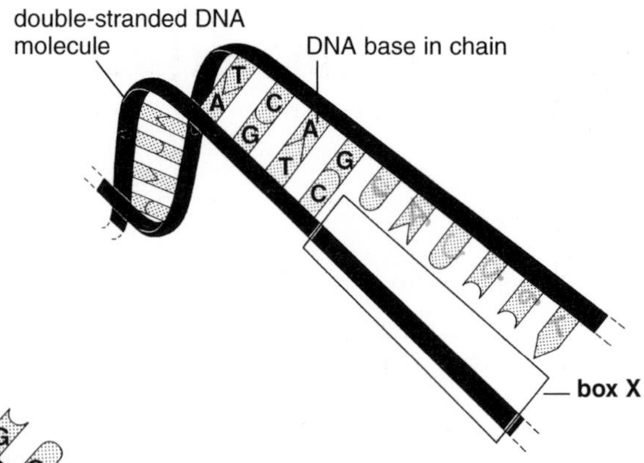

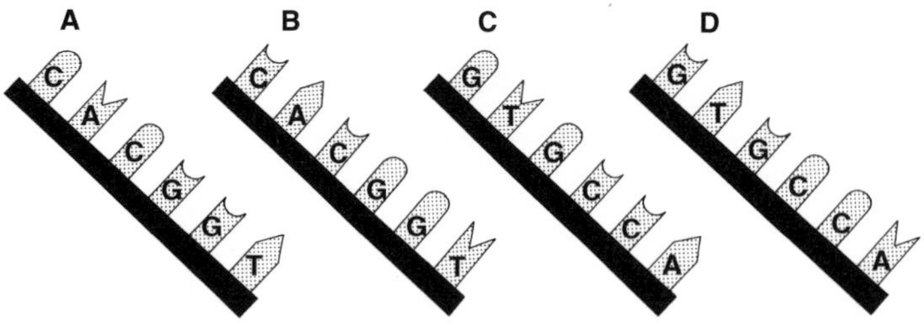

Which part of the second diagram supplies the information missing from **box X** in the first diagram?

8 The only difference between the DNA of one member of a species and that of another member of the same species is the

A order in which the bases occur in their chromosomes.
B type of bases present in their chromosomes.
C number of strands that occur in their chromosomes.
D type of bonds present between the bases in their chromosomes.

9 How many DNA bases make up a 'codeword' for an amino acid?

A 1 **B** 2 **C** 3 **D** 4

10 The production of a molecule of protein involves the stages in the following list.

1 messenger molecule passes out of nucleus into cytoplasm
2 region of DNA in nucleus opens up
3 protein is made according to the order of codewords on messenger molecule
4 messenger molecule is made against a DNA strand

The correct order of these stages is

A 2, 4, 3, 1. **B** 2, 4, 1, 3. **C** 4, 2, 1, 3. **D** 4, 2, 3, 1.

11 The number of matching sets of chromosomes present in a normal body cell of a human being is

A 2. **B** 4. **C** 23. **D** 46.

Questions 12, 13 and 14 refer to the accompanying diagram which represents the human life cycle.

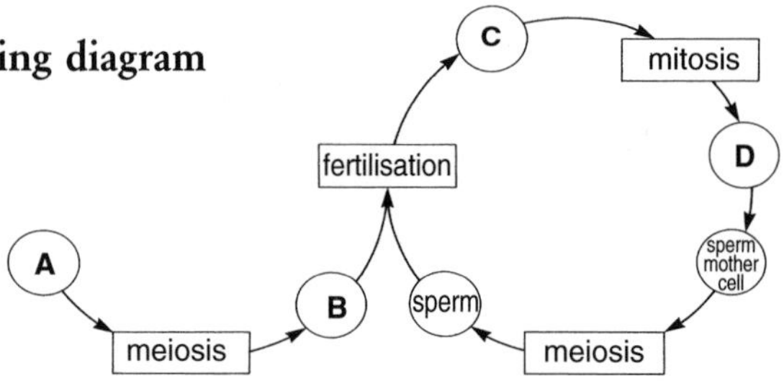

12 Which letter represents an egg?

13 Which letter represents a zygote?

14 Which letter represents an egg mother cell?

Items 15, 16, 17, 18 and 19 refer to the accompanying diagram of a gamete mother cell before cell division.

15 The number of chromosomes present in this cell is

A 2. **B** 4. **C** 8. **D** 16.

16 The number of complete sets of chromosomes present in this cell is

A 2. **B** 4. **C** 8. **D** 16.

17 The number of chromosome pairs present in this cell is

A 2. **B** 4. **C** 8. **D** 16.

18 The number of chromatids present in this cell is

A 4. **B** 8. **C** 16. **D** 32.

19 The number of chromosomes present in each gamete produced by this cell would be

A 2. **B** 4. **C** 6. **D** 8.

20 The first of the two accompanying diagrams shows a set of four gametes. (The paternal chromosomes are shaded black and the maternal chromosomes are unshaded.)

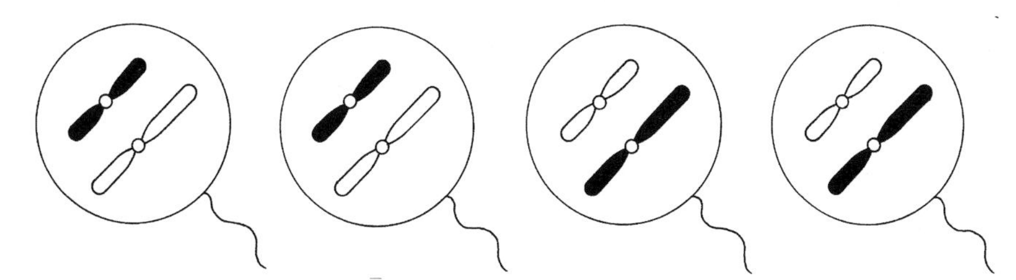

Which part of the second diagram represents the gamete mother cell that produced the four gametes?

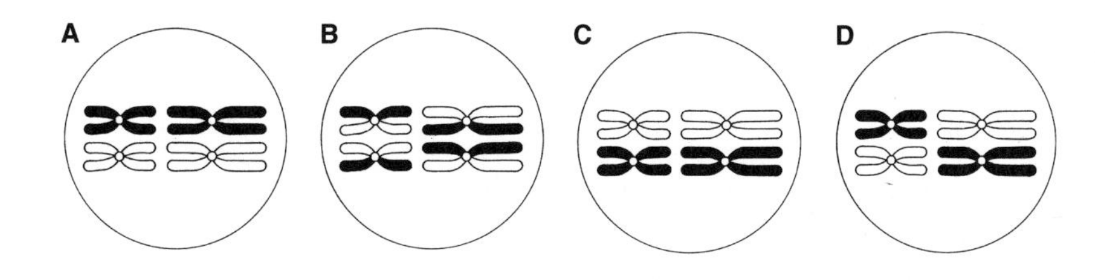

21 The number of different combinations of chromosomes that can be produced in the gametes of gamete mother cells containing THREE pairs of chromosomes is

A 3. **B** 4. **C** 6. **D** 8.

22 Random assortment of chromosomes during gamete formation is important because it

A leads to variety amongst the sex cells formed.
B maintains the chromosome number from cell to cell.
C ensures that the genetic instructions are copied exactly.
D produces mutant chromosomes different from those in the parent cell.

23 During sperm production in a human male, the chance of a sperm receiving a Y chromosome is

A 1 in 1. **B** 1 in 2.
C 1 in 3. **D** 1 in 4.

Items 24 and 25 refer to the accompanying diagram which shows the chromosome number present in the body cells of a species of bird. X and Y are the sex chromosomes.

These two items also refer to the possible answers contained in the accompanying boxed diagram.

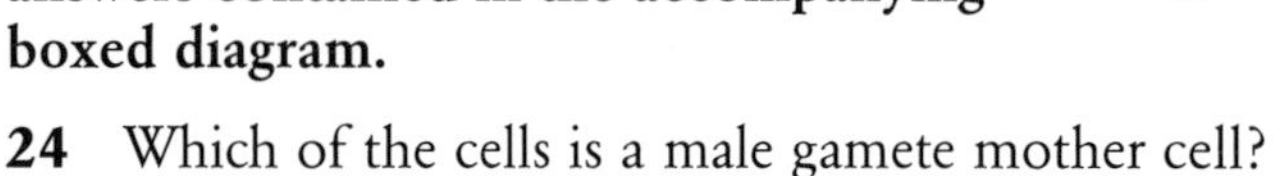

24 Which of the cells is a male gamete mother cell?

A 1 **B** 2 **C** 3 **D** 4

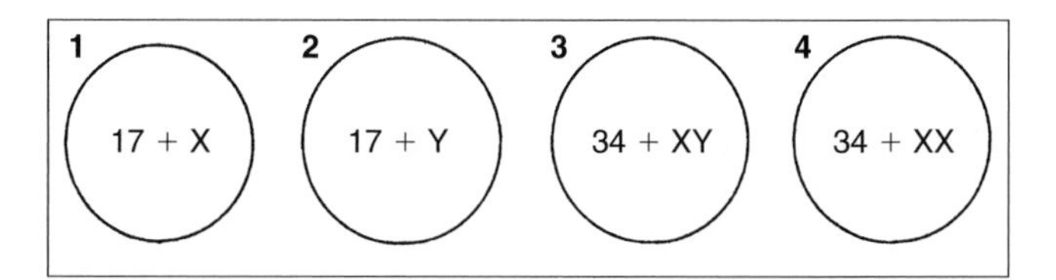

25 Eggs are correctly depicted by

A 1 only. **B** 2 only.
C 1 and 2. **D** 3 and 4.

Test 2

1 In a human female, fertilisation normally takes place in the

A vagina. **B** uterus. **C** oviduct. **D** ovary.

2 On average, the diameter of a human egg is 120 μm. Expressed as a fraction of a millimetre, this would be

A 0.0012 **B** 0.012 **C** 0.12 **D** 1.2

Items 3 and 4 refer to the accompanying diagram of the human male reproductive system.

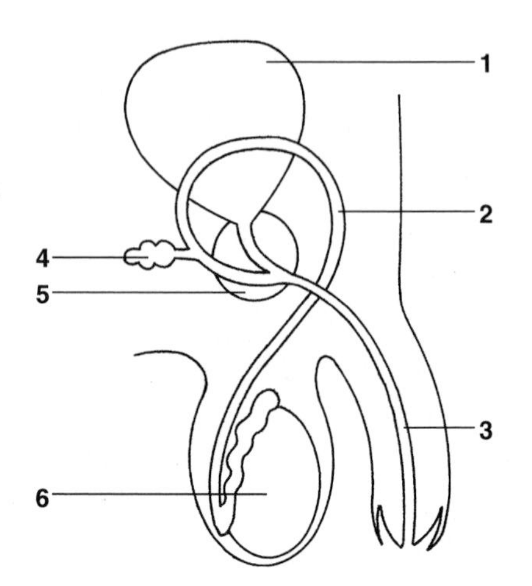

3 Sperm are produced in

A 1. **B** 4. **C** 5. **D** 6.

4 The routine male sterilisation operation known as vasectomy is carried out by

A collapsing 1. **B** cutting 2.
C blocking 3. **D** removing 6.

Items 5 and 6 refer to the accompanying bar graph.

5 The average number of eggs produced annually by a curlew is

A 3. **B** 4. **C** 5. **D** 6.

6 Which bar in the graph represents the robin which produces, on average, 13 eggs per year?

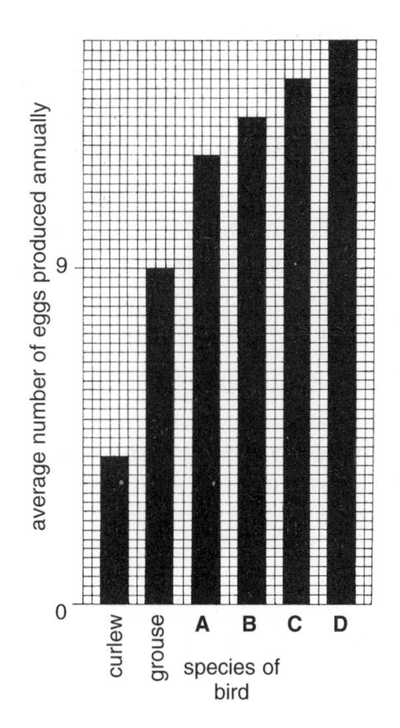

Items 7 and 8 refer to the accompanying diagram which shows part of a flower.

7 Pollination is the transfer of pollen from

A 1 to 2. **B** 2 to 3.
C 4 to 5. **D** 1 to 6.

8 Fertilisation takes place at point

A 1. **B** 2. **C** 3. **D** 5.

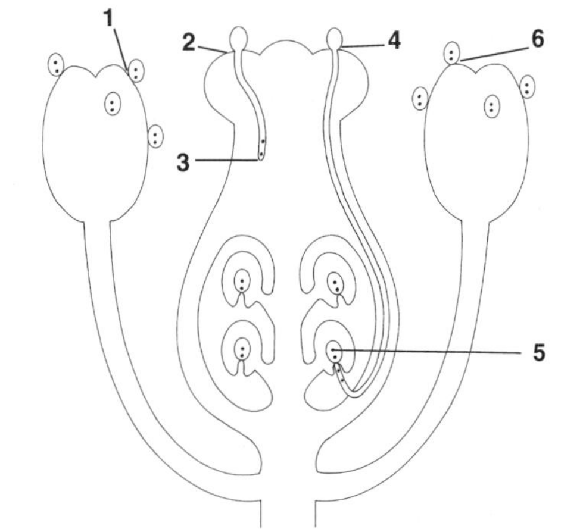

Items 9, 10 and 11 refer to the accompanying diagram of part of a cell's genetic material.

9 Which boxed region represents a complete gene?

10 Which boxed region represents a DNA base?

11 Which boxed region represents a chromosome?

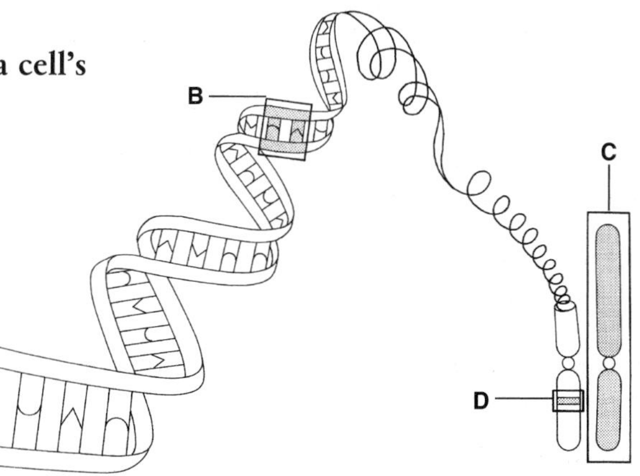

12 How many different types of amino acid are found to occur in proteins?

A 3 **B** 4 **C** 20 **D** 64

13 The sequence of amino acids in a protein is determined indirectly by the order of the

A bases in a region of DNA molecule.
B genes in a region of enzyme molecule.
C bases in a region of polypeptide chain.
D genes in a region of polypeptide chain.

14 Which line in the table is CORRECT?

	Number of complete sets of chromosomes in cell	
	Gamete mother cell	**Gamete**
A	1	1
B	1	2
C	2	1
D	2	2

15 The accompanying diagram represents the human life cycle.

Which line in the following table correctly matches the numbered boxes in the diagram?

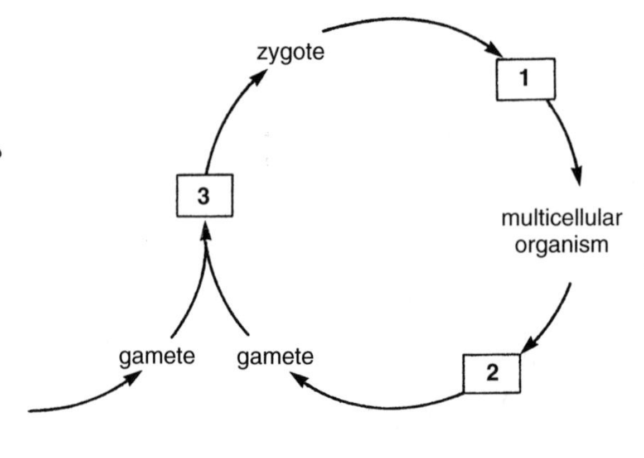

	1	2	3
A	mitosis	fertilisation	meiosis
B	meiosis	mitosis	fertilisation
C	fertilisation	meiosis	mitosis
D	mitosis	meiosis	fertilisation

16 Which of the following could be used to prepare a microscope slide of cells undergoing meiosis?

A locust testis
B pea root tip
C human cheek epithelium
D fruit fly salivary gland

Questions 17 and 18 refer to the information in the following list.

1 23 chromosomes
2 23 pairs of chromosomes
3 46 chromosomes
4 46 pairs of chromosomes

17 Which numbered entry on the list refers to the genetic content of a human gamete?

A 1 **B** 2 **C** 3 **D** 4

18 Which two entries on the list BOTH refer to the genetic content of the gamete mother cells in a human male?

A 1 and 2 **B** 1 and 3
C 2 and 3 **D** 2 and 4

19 The accompanying diagram shows four sets of gametes being produced by two gamete mother cells. (The paternal chromosomes are shaded black and the maternal chromosomes are unshaded.) Which set of gametes is CORRECT?

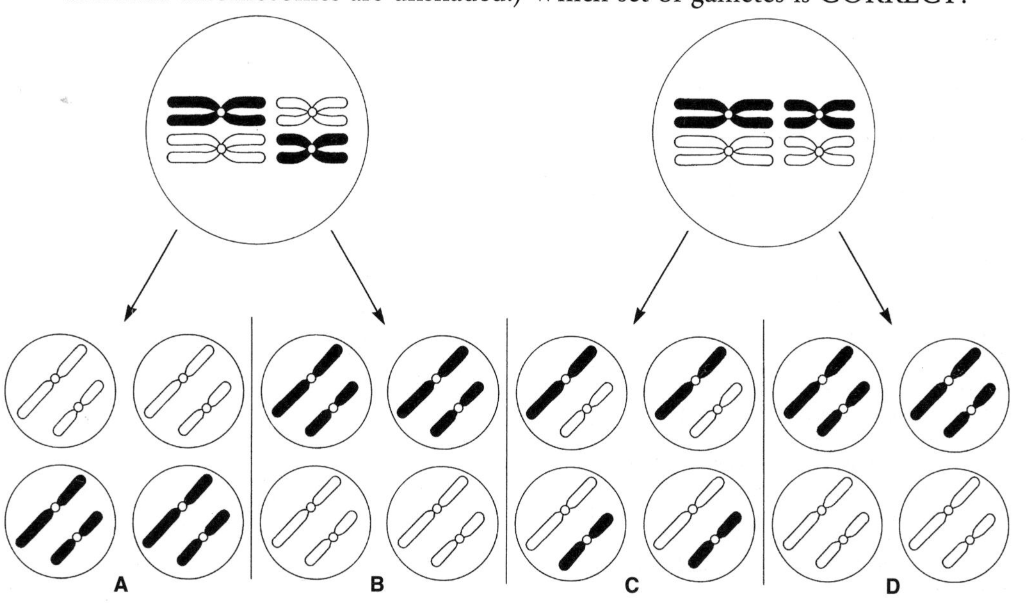

20 The accompanying diagram shows 7 of the 8 different types of gamete that can result from random assortment of 3 pairs of matching chromosomes present in an organism's gamete mother cells.

Which of the following is the eighth gamete?

 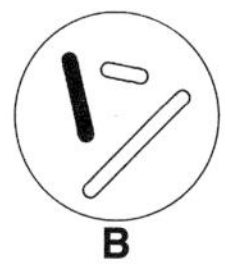  

A B C D

21 The first of the two accompanying tables refers to the chromosome numbers of several cells but it is incomplete.

Which line in the second table provides the missing information?

Chromosome number		Number of different combinations of matching chromosomes that can arise in gametes following random assortment
In body cell	**In gamete**	
4	**Y**	$2^2 = 4$
6	3	**Z**
X	4	$2^4 = 16$

	X	Y	Z
A	8	1	$2^3 = 6$
B	8	2	$2^3 = 8$
C	10	1	$2^3 = 6$
D	10	2	$2^3 = 8$

22 It is important that sexual reproduction produces variation amongst the members of a species so that the species

A is able to control its fertility rate.
B can adapt to a changing environment.
C is prevented from reproducing asexually.
D can alter its chromosome complement from generation to generation.

23 Which of the following represents the total number of chromosomes present in a normal body cell of a human male?

A 44 + XX **B** 44 + XY **C** 46 + XX **D** 46 + XY

24 During egg production in a human female, the chance of an egg receiving an X chromosome is

A 1 in 1. **B** 1 in 2. **C** 1 in 3. **D** 1 in 4.

25 Which of the following statements refers CORRECTLY to the sperm cells present in a normal sample of human semen?

A All of them contain an X chromosome.
B All of them contain a Y chromosome.
C Some of them contain both an X and a Y chromosome.
D Some of them contain an X and some contain a Y chromosome.

Genotype and phenotype

Test 1

Items 1 and 2 refer to the following possible answers.

A allele **B** gamete **C** genotype **D** phenotype

1 Which term means the total set of genes possessed by an organism?

2 Which term means the physical appearance of an organism?

3 The number of alleles of each gene present in a gamete of an animal is

A 1. **B** 2. **C** 23. **D** 46.

Items 4, 5, 6 and 7 refer to the accompanying information about a genetics experiment involving tobacco plants.

[**J**] true-breeding green plants × true-breeding white plants

↓

[**K**] all green plants

offspring from first cross self-pollinated — green plants × green plants

↓

[**L**] green plants and white plants

4 The symbols that should have been used at positions **J**, **K** and **L** are

	J	K	L
A	F_1	P	F_2
B	P	F_1	F_2
C	P_1	F_1	P_2
D	F_1	P_1	P_2

5 The term used to refer to a cross where the original parents differ from one another in one way only is

A filial. **B** monohybrid. **C** homozygous. **D** true-breeding.

6 With reference to leaf colour in tobacco plants, it is correct to say that

A green and white are co-dominant.
B white and green are co-recessive.
C green is dominant and white is recessive.
D white is dominant and green is recessive.

7 The phenotypic ratio shown by the members of the generation marked **[L]** would be

A 1 white : 1 green. **B** 3 white : 1 green.
C 3 green : 1 white. **D** 4 green : 0 white.

8 The cross in the accompanying diagram involves the gene for coat colour in cattle.

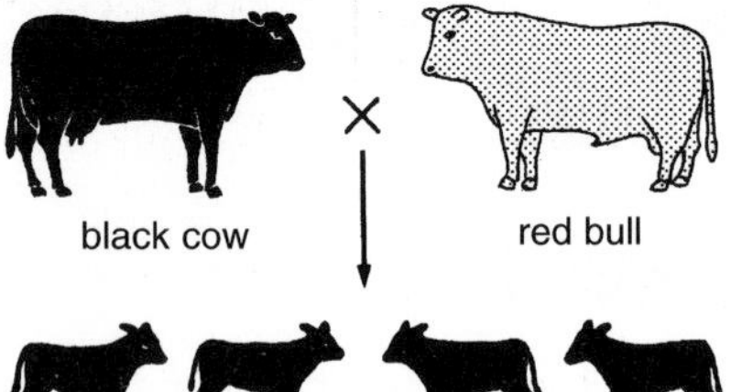

Which of the following keyed diagrams CORRECTLY represents this cross?

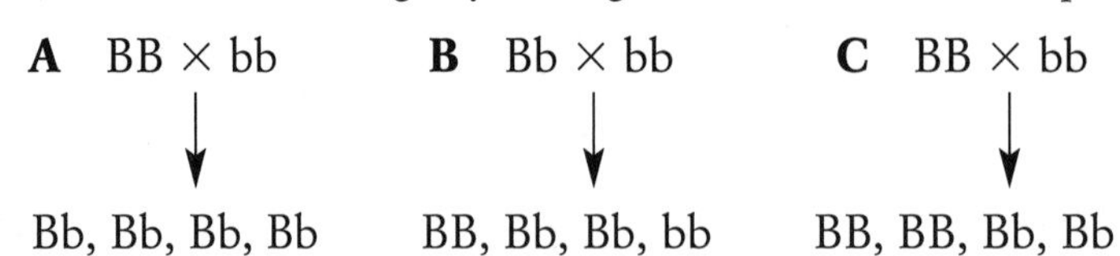

9 The Punnett square in the accompanying diagram shows the results of crossing organisms with the genotype Rr, where R is the dominant allele and r the recessive allele of a gene.

Which TWO boxes represent organisms that would have the same phenotype but different genotypes?

male gametes

female gametes	R	r
R	①	②
r	③	④

A 1 and 2 **B** 2 and 3 **C** 1 and 4 **D** 3 and 4

10 In guinea pigs, long hair is recessive to short hair. If a large group of long-haired females is crossed with a large group of heterozygous males, the percentage of their offspring that are long-haired will be approximately

A 25. **B** 50. **C** 75. **D** 100.

11 In gerbils, brown coat (B) is dominant to grey coat (b). If a gerbil has the genotype Bb, then its parents could NOT have been

A BB and Bb. **B** bb and BB.
C BB and BB. **D** Bb and Bb.

12 In pea plants, the allele for flower colour (H) is dominant to the allele for lack of flower colour (h). A plant homozygous for flower colour was crossed with a plant bearing colourless flowers. The F_1 plants were then self-pollinated.

Which of the following correctly represents the ratio of genotypes expected in the F_2 generation?

A all Hh **B** 1 HH : 1 hh
C 3 HH : 1 hh **D** 1 HH : 2 Hh : 1 hh

13 In fruit flies, the allele for normal grey body colour (G) is dominant to the allele for ebony body colour (g). The accompanying table summarises the results of several crosses.

Which strains BOTH have the genotype Gg?

A 1 and 3 **B** 1 and 4
C 2 and 3 **D** 2 and 4

Cross	Result
strain 1 × gg	all grey
strain 2 × gg	1 grey : 1 ebony
strain 3 × gg	all ebony
strain 4 × Gg	3 grey : 1 ebony

14 In fruit flies, long wing is dominant to vestigial (tiny) wing. When heterozygous long-winged flies were crossed with vestigial-winged flies, 192 offspring were produced. Of these, 101 had long wings and 91 had vestigial wings. If an exact ratio had been obtained, then the number of each phenotype would have been.

	Long-winged	Vestigial-winged
A	64	128
B	96	96
C	128	64
D	192	0

Items 15, 16 and 17 refer to the accompanying diagram of a family tree and to the following information. In humans, the allele for red hair (h) is recessive to the allele for non-red hair (H).

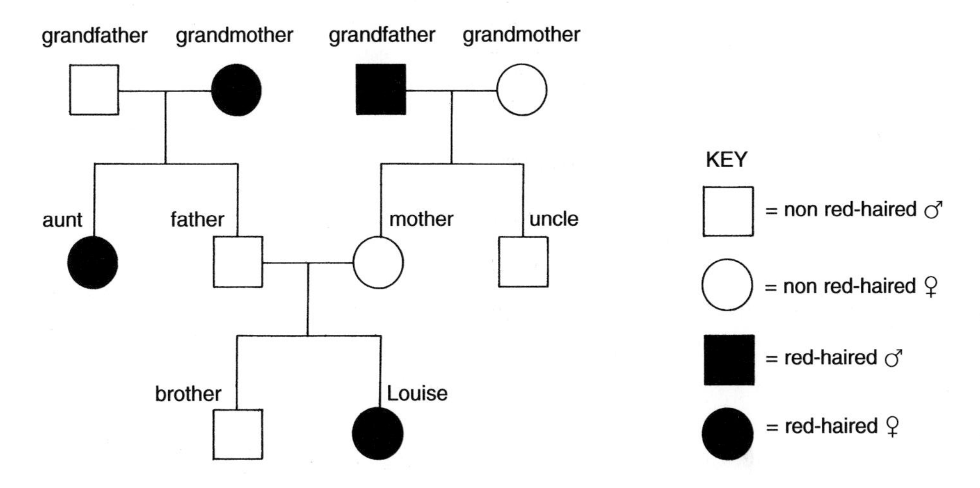

15 Which line in the following table is CORRECT?

	Genotype of Louise	Genotype of Louise's parents
A	homozygous	heterozygous
B	homozygous	homozygous
C	heterozygous	homozygous
D	heterozygous	heterozygous

16 Which line in the following table CORRECTLY identifies the genotypes of Louise's aunt and uncle?

	Aunt	Uncle
A	hh	HH
B	Hh	hh
C	hh	Hh
D	HH	Hh

17 Each time Louise's parents produce a child, the chance of it being red-haired is

A 1 in 1. **B** 1 in 2.
C 1 in 3. **D** 1 in 4.

Items 18, 19 and 20 refer to the accompanying diagram of a family tree where the allele for wavy hair (W) is dominant to the allele for straight hair (w).

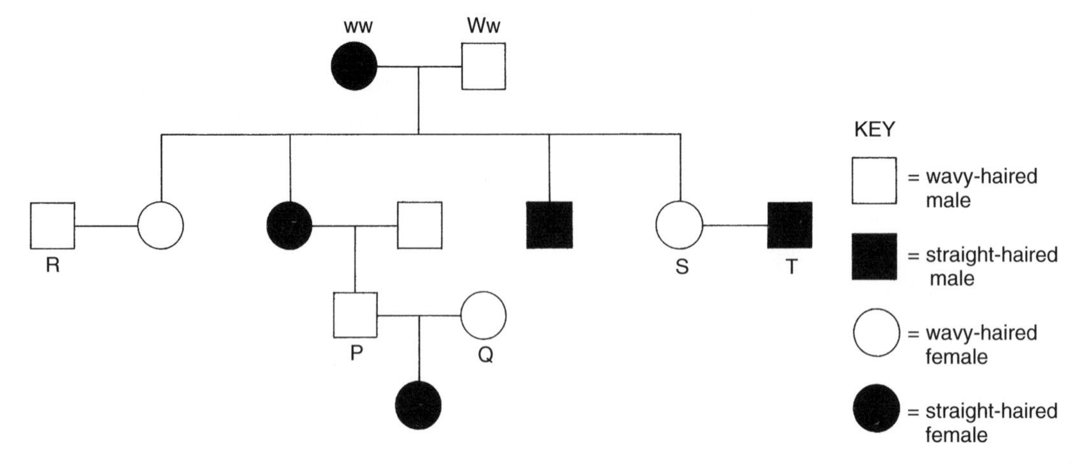

Items 18 and 19 also refer to the following possible answers.

A They have the same phenotype and the same genotype.
B They differ in phenotype but have the same genotype.
C They differ in both phenotype and genotype.
D They have the same phenotype but differ in genotype.

18 If person R is homozygous for the wavy hair allele, then which statement is true about persons R and S?

19 Which statement is true about persons P and Q?

20 The chance of persons S and T producing a straight-haired child is

A 1 in 1. **B** 1 in 2. **C** 1 in 3. **D** 1 in 4.

Items 21 and 22 refer to the family tree shown in the accompanying diagram.

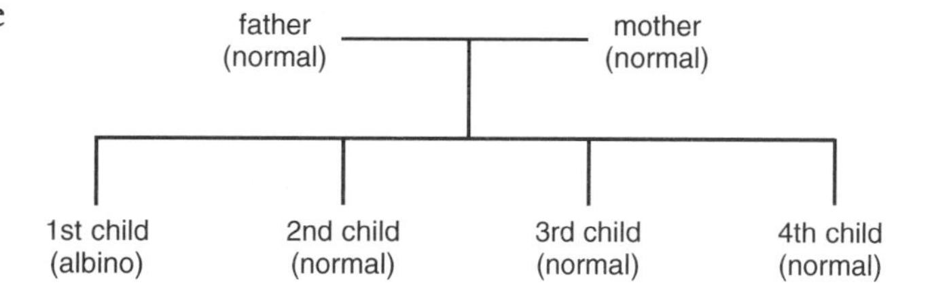

21 If A = normal allele and a = albino allele, the genotypes of the parents in the family tree are

	Father	Mother
A	AA	AA
B	AA	Aa
C	Aa	AA
D	Aa	Aa

22 The chance of this couple's fifth child being an albino is

A 1 in 2. **B** 1 in 3. **C** 1 in 4. **D** 1 in 5.

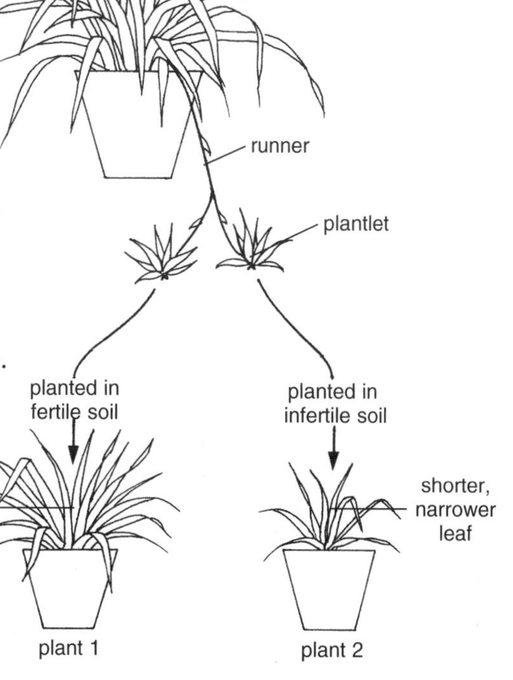

23 The accompanying diagram shows the outcome of transplanting plantlets from a spider plant to two types of soil.

It is CORRECT to say that plants 1 and 2 have

A the same genotype but different phenotypes.
B the same genotype and the same phenotype.
C different genotypes but the same phenotype.
D different genotypes and different phenotypes.

Questions 24 and 25 refer to the data in the accompanying table which were obtained from studies of pairs of twins.

Characteristic measured	1	2	3	4
	50 pairs of identical twins reared together	50 pairs of non-identical twins reared together	50 pairs of identical twins reared apart	50 pairs of non-identical twins reared apart
Average difference in body mass (kg)	1.8	4.5	4.4	4.6
Average difference in width of head (mm)	2.8	4.3	2.9	4.4
Average difference in waist circumference (cm)	2.0	9.8	9.5	10.4
Average number of fillings needed to teeth	2.3	8.7	10.6	9.9

24 Which two columns should be compared to find out if the environment has an effect when the genotype is the same?

A 1 and 2 **B** 1 and 3
C 2 and 3 **D** 3 and 4

25 Which characteristic appears to be controlled mainly by genes?

A body mass
B width of head
C waist circumference
D number of fillings

Test 2

Items 1 and 2 refer to the following possible answers

A dominant **B** heterozygous
C homozygous **D** recessive

1 Which term describes a genotype that contains two different alleles of a particular gene?

2 Which term describes the member of a pair of alleles that is always expressed in the phenotype?

3 How many alleles of each gene does a zygote receive from each parent at fertilisation?

A 1 **B** 2 **C** 23 **D** 46

Items 4 and 5 refer to the following information. In pea plants, the gene for height has two alleles, tall and dwarf. The cross shown in the accompanying diagram was carried out and then a further generation of pea plants was produced by allowing the first filial generation to self-pollinate.

4 Which of the following Punnett squares represents the second cross?

A

		genotypes of pollen	
		T	T
genotypes of eggs	t	tT	tT
	t	tT	tT

B

		genotypes of pollen	
		T	t
genotypes of eggs	T	TT	Tt
	t	tT	tt

C

		genotypes of pollen	
		T	t
genotypes of eggs	T	TT	TT
	t	tt	tt

D

		genotypes of pollen	
		t	t
genotypes of eggs	T	Tt	Tt
	T	Tt	Tt

parents (both true-breeding)

tall × dwarf

first filial generation

all tall

5 The phenotypic ratio of tall to dwarf amongst the offspring of the second cross would be

A 1:1. **B** 2:1. **C** 3:1. **D** 4:1.

6 In mice, black coat colour (allele B) is dominant to brown coat colour (allele b). The offspring of a cross between a black mouse (BB) and a brown mouse were allowed to interbreed. What percentage of the progeny would be expected to have black coats?

A 25 **B** 50 **C** 75 **D** 100

Items 7, 8 and 9 refer to the following information about crosses involving tomato plants.

cross 1 hairy plant × smooth plant

↓

F_1 all hairy plants

cross 2 F_1 hairy plant × F_1 hairy plant

↓

F_2 hairy plants and smooth plants

7 The number of different genotypes present in the F_1 generation is

A 1. **B** 2. **C** 3. **D** 4.

8 The number of different genotypes present in the F_2 generation is

A 1. **B** 2. **C** 3. **D** 4.

9 If F_1 hairy plants are crossed with smooth plants, then a possible result for the offspring could be

A 1000 hairy and 0 smooth.
B 756 hairy and 244 smooth.
C 666 hairy and 334 smooth.
D 492 hairy and 508 smooth.

10 In tobacco plants, the gene for leaf colour has two alleles, green and white, where green (G) is dominant to white (g). If a heterozygous green plant is crossed with a white plant, the genotypes of the offspring produced will be

A all Gg. **B** Gg and gg.
C GG and gg. **D** GG, Gg and gg.

11 Various crosses involving three red-coloured plants (1, 2 and 3) and one white-coloured plant (4) were carried out as shown in the accompanying table.

Cross	Offspring produced	
	red-flowered (%)	white-flowered (%)
1 × 4	100	0
2 × 4	50	50
2 × 3	75	25

Which of the following plants have the same genotype?

A 1 and 2 **B** 1 and 3 **C** 2 and 3 **D** 1, 2 and 3

Questions 12 and 13 refer to the accompanying diagram which shows a series of crosses carried out using leopards.

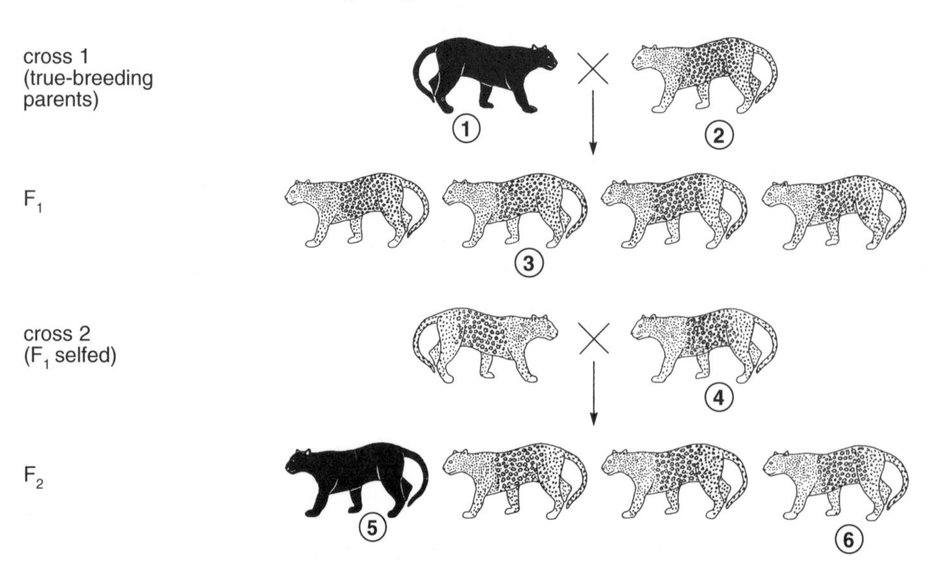

12 Which animal is definitely heterozygous with respect to the gene for coat type?

A 1 **B** 2 **C** 3 **D** 5

13 Which animal is definitely homozygous with respect to the gene for coat type?

A 3 **B** 4 **C** 5 **D** 6

14 In humans, there are two alleles of the gene which controls the ability to taste a chemical called phenylthiocarbamide. Ability to taste the chemical (T) is dominant over inability to taste it (t).

The accompanying diagram shows the inheritance of this characteristic amongst the members of a family.

Which line in the following table identifies the three genotypes missing from the family tree?

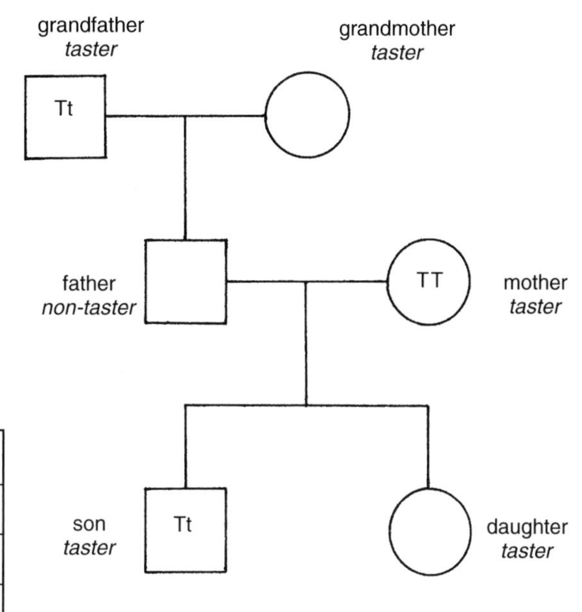

	Grandmother	Father	Daughter
A	TT	Tt	TT
B	Tt	tt	TT
C	TT	Tt	Tt
D	Tt	tt	Tt

Items 15, 16 and 17 refer to the accompanying diagram of a family tree and to the following information. The ability to roll the tongue is governed by the presence of the dominant allele R. The recessive allele for inability to roll the tongue is r.

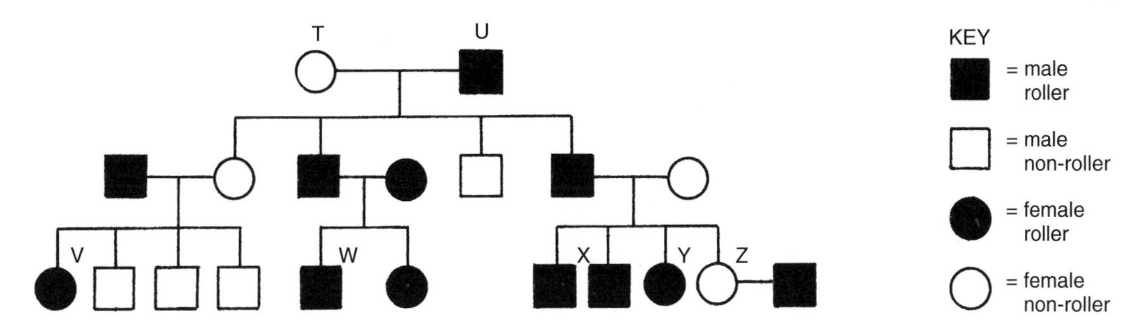

15 How many male grandchildren do grandparents T and U have?

A 3 **B** 6 **C** 7 **D** 11

16 Which person could have the genotype RR?

A V **B** W **C** X **D** Y

17 Person Z marries a person of genotype Rr. The chance of each of their children being a tongue roller is

A 1 in 1. **B** 1 in 2. **C** 1 in 3. **D** 1 in 4.

18 In humans, wavy hair is dominant to straight hair. A wavy-haired woman marries a straight-haired man and they have four children. One son and one daughter both have wavy hair and one son and one daughter both have straight hair. Which of the following family trees CORRECTLY represents this information?

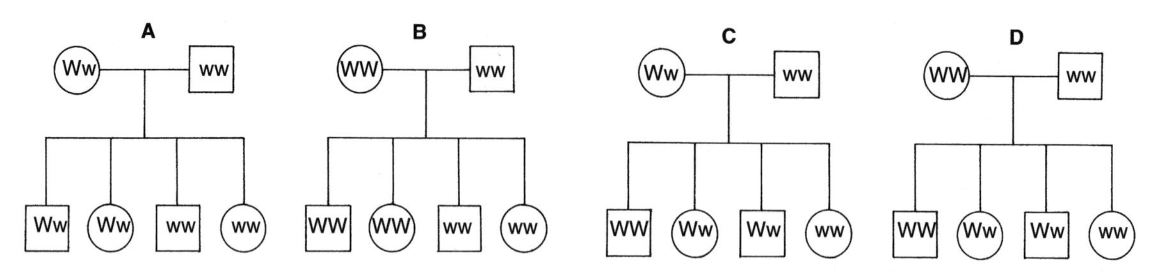

19 Allele O is recessive to both A and B, the co-dominant alleles of the gene determining human blood group. In which of the following paternity suits is the alleged father definitely NOT the child's father?

Paternity suit	Mother's genotype	Alleged father's genotype	Child's blood group
A	BO	AA	AB
B	AO	BO	O
C	OO	AB	A
D	AO	OO	B

20 The accompanying diagram shows the outcome of transplanting plantlets 1 and 2 from a spider plant to fertile soil and then keeping them in different conditions of light.

What would be the result of transplanting plantlets 3 and 4 to fertile soil and keeping both of them in bright light for several weeks?

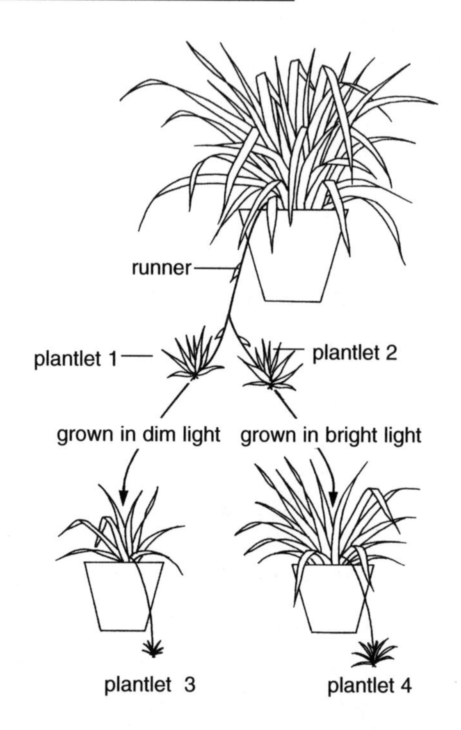

	Plantlet 3	Plantlet 4
A	small unhealthy plant	small unhealthy plant
B	small unhealthy plant	medium-sized healthy plant
C	medium-sized healthy plant	small unhealthy plant
D	medium-sized healthy plant	medium-sized healthy plant

21 Which of the following does NOT form part of the theory of evolution proposed by Charles Darwin?

A Any beneficial change in an organism's phenotype is brought about by the direct action of the environment.
B Members of the same species are not identical but show variation in all characteristics.
C A struggle for existence occurs because organisms tend to produce more offspring than the environment will support.
D Those offspring whose phenotypes are less well suited to the environment are more likely to die before producing offspring.

Items 22, 23, 24 and 25 refer to the following information. The peppered moth exists in two forms: the light-coloured variety and the dark (melanic) type. In an experiment, individuals of both types were marked on their undersides with a dot of paint and then some were released in a rural area and some were released in an industrial area.

Many of these marked moths were later recaptured, as shown in the following table.

	Rural area		Industrial area	
	light moth	**melanic moth**	**light moth**	**melanic moth**
Number of marked moths released	250	200	250	300
Number of marked moths recaptured	40	**see question 23**	45	162
Percentage number of marked moths recaptured	16	4	18	54

22 Melanic moths occur as a result of

A industrial pollution. **B** natural selection. **C** evolution. **D** mutation.

23 How many melanic moths were recaptured in the rural area?

A 2 **B** 4 **C** 8 **D** 20

24 From the data in the table, it is NOT valid to conclude that

A in the rural area, light moths were four times more likely to survive than melanic moths.
B a greater percentage number of both types of moth were recaptured in the industrial area compared with the rural area.
C in the industrial area, melanic moths were three times more likely to survive than light-coloured moths.
D the total percentage number of light moths recaptured in both areas exceeded the total percentage number of melanic moths recaptured.

25 The melanic moth enjoys a selective advantage in an industrial area because

A predators fail to notice it against a sooty background.
B there is no competition since the light form is killed by pollution.
C predators ignore it because it is dirty and noxious to eat.
D it is easily seen against light-coloured tree trunks.

10 Applied genetics

Test 1

1 The accompanying diagram shows a simplified version of selective breeding in cattle. Which line in the table correctly describes the ancestor from which modern cattle have arisen?

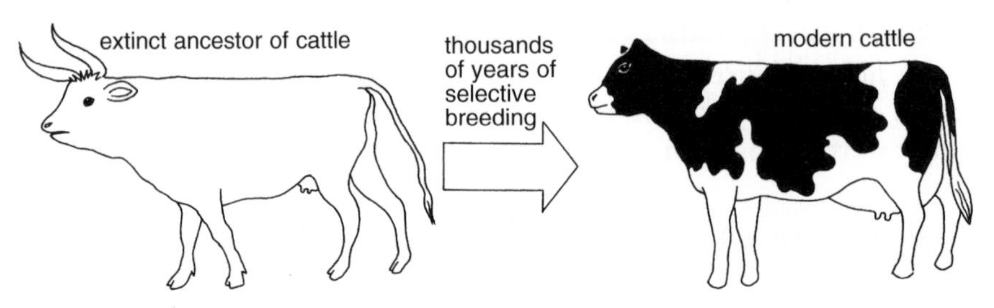

	Size of milk sacs		Horns		Rate of movement	
	Large	Small	Present	Absent	Fast	Slow
A	✗	✓	✓	✗	✓	✗
B	✓	✗	✗	✓	✓	✗
C	✓	✗	✗	✓	✗	✓
D	✓	✗	✓	✗	✗	✓

2 The following table refers to five varieties of pea plant.

Characteristic	Variety of pea plant				
	1	2	3	4	5
Ability to form root nodules	++	+++	++	+	++
Ability to resist drought	+++	+	+++	++	+
Ability to resist pests	++	++	+	+++	++
Ability to survive global warming	+	++	+++	+	+

+++ = high ++ = medium + = low

Which variety of pea plant should be crossed with variety 5 in an attempt to produce an enhanced strain with a rating of medium or high for all four characteristics?

A 1 **B** 2 **C** 3 **D** 4

Items 3 and 4 refer to the following table which gives some of the characteristics of four breeds of cattle.

Breed	Milk yield	Meat yield	Incidence of twin calves	Hide quality
1	medium	medium	1 in 3	excellent
2	high	low	1 in 2	good
3	medium	high	1 in 5	good
4	high	low	1 in 10	excellent

3 A farmer wishes to improve the milk yield of his cattle and increase their number. Which breed of bull should he introduce to his cattle?

A 1 **B** 2 **C** 3 **D** 4

4 Which two breeds of cattle should be crossed in an attempt to produce a new breed with enhanced meat yield and hide quality?

A 1 and 2 **B** 1 and 3
C 2 and 3 **D** 2 and 4

Items 5, 6 and 7 refer to the accompanying graph which shows the results of selectively breeding 50 generations of maize plants for increased percentage protein content of the grains.

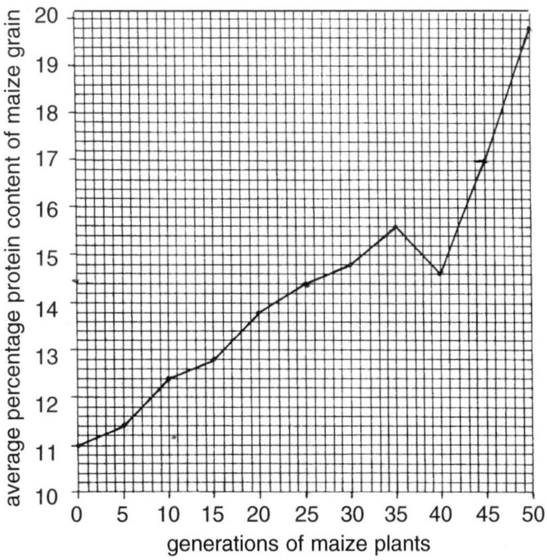

5 After 25 generations of selective breeding, the average percentage protein content of maize grains was

A 14.2. **B** 14.4.
C 14.6. **D** 14.8.

6 Between which generations did the greatest increase in percentage protein content of maize grains occur?

A 5–10 **B** 30–35 **C** 40–45 **D** 45–50

7 From generation 10 to generation 30, the average increase in percentage protein content per generation is

A 0.12. **B** 0.24. **C** 1.20. **D** 2.40.

Items 8, 9 and 10 refer to the accompanying diagram which shows the genetic material of a bacterium.

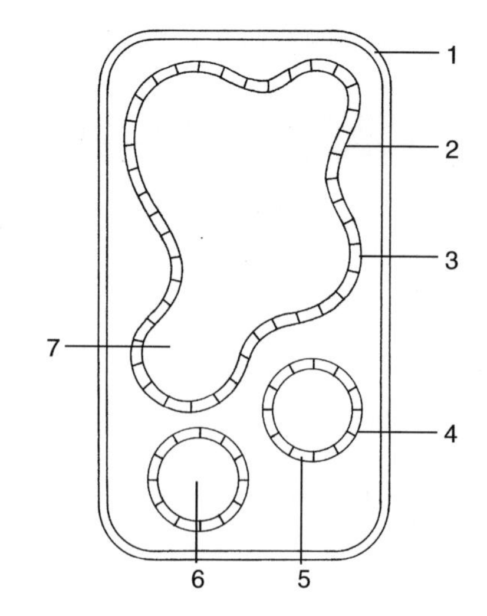

8 Which structure is the chromosome?

A 1 **B** 2 **C** 4 **D** 6

9 Which structure is a plasmid?

A 2 **B** 3 **C** 4 **D** 7

10 Which structures are BOTH genes?

A 2 and 3 **B** 3 and 4
C 3 and 5 **D** 5 and 6

Items 11, 12 and 13 refer to the following list of procedural steps employed during a genetic engineering experiment.

1 host cell allowed to multiply
2 required DNA fragment cut out of appropriate chromosome
3 duplicate plasmids formed which express 'foreign' gene
4 plasmid extracted from bacterium and opened up
5 altered plasmid inserted into bacterial host cell
6 DNA fragment sealed into plasmid

11 The correct order in which these steps would be carried out is

A 2, 4, 6, 5, 1, 3. **B** 2, 4, 5, 6, 3, 1.

C 4, 6, 2, 5, 3, 1. **D** 4, 6, 2, 5, 1, 3.

12 A special enzyme that acts as biochemical 'scissors' would be used during steps

A 2 and 4. **B** 2 and 6.
C 4 and 5. **D** 4 and 6.

13 A special enzyme that acts as biochemical 'glue' would be used at stage

A 2. **B** 4. **C** 5. **D** 6.

14 Insulin produced by genetic engineering is given to people who would otherwise suffer

A haemophilia.
B reduced growth.
C diabetes mellitus.
D internal blood clotting.

15 Which line in the following table is CORRECT?

	Transgenic plant	Role of inserted gene	Beneficial effect
A	strawberry	production of anti-freeze chemical	leaves resist attack by caterpillars
B	pea	production of insecticide protein	crop survives but weeds die when weedkiller is applied
C	soya	production of protein that gives resistance to weedkiller	seeds are protected against damage by frost
D	tomato	blockage of production of an enzyme	fruit is prevented from becoming soft and mushy

Test 2

1 Which part of the accompanying diagram represents selective breeding in pigs?

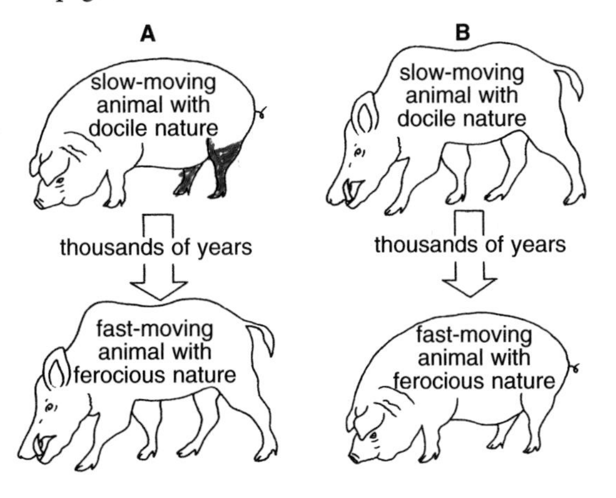

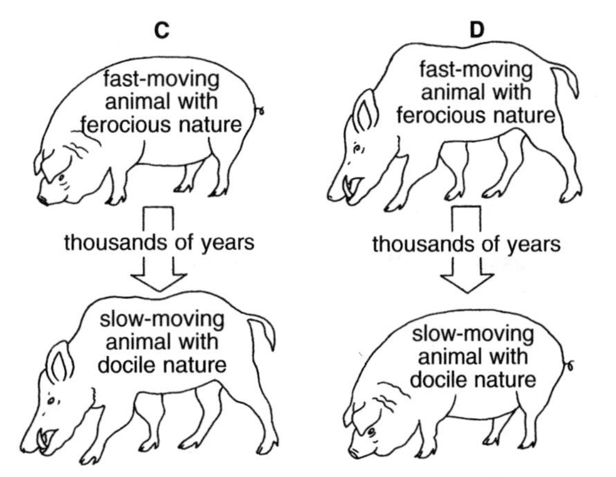

2 The data in the accompanying table show the effect that selective breeding had on the milk yield of Ayrshire cattle over a period of time.

The percentage increase in milk yield for the period shown is

A 0.5. **B** 5.0. **C** 5.5. **D** 20.0.

Year	Annual milk yield (l)
1	3280
2	3320
3	3362
4	3401
5	3444

Questions 3, 4 and 5 refer to the accompanying table which gives the results from an experiment to investigate the effect of selective breeding on the percentage oil content of a species of cereal grain.

Number of generations of selective breeding	Average percentage of oil in grain	
	Strain X (selected for most oil)	Strain Y (selected for least oil)
start	4.5	4.5
10	7.3	2.7
20	8.1	2.4
30	10.2	1.8
40	9.5	1.1
50	15.0	0.9

3 Between which of the following generations did strain X show an increase in percentage oil content of 0.8?

A 10 to 20 **B** 20 to 30 **C** 30 to 40 **D** 40 to 50

4 Between which of the following generations did strain X FAIL to show an increase in percentage oil content of grains?

A 10 to 20 **B** 20 to 30 **C** 30 to 40 **D** 40 to 50

5 Between which of the following generations did strain Y show the greatest decrease in percentage oil content of grains?

A 10 to 20 **B** 20 to 30 **C** 30 to 40 **D** 40 to 50

Items 6, 7, 8 and 9 refer to the accompanying graph which shows the effects of selective breeding on the milk yield and percentage butterfat produced by a certain breed of cattle over a 30-year period.

6 During which year was the milk yield found to be 10 tonnes and the butterfat content 3.6%?

A 1920 **B** 1925
C 1930 **D** 1935

7 During which 5-year interval of time did the greatest increase in percentage butterfat content of milk occur?

A 1910–1915 **B** 1920–1925
C 1925–1930 **D** 1935–1940

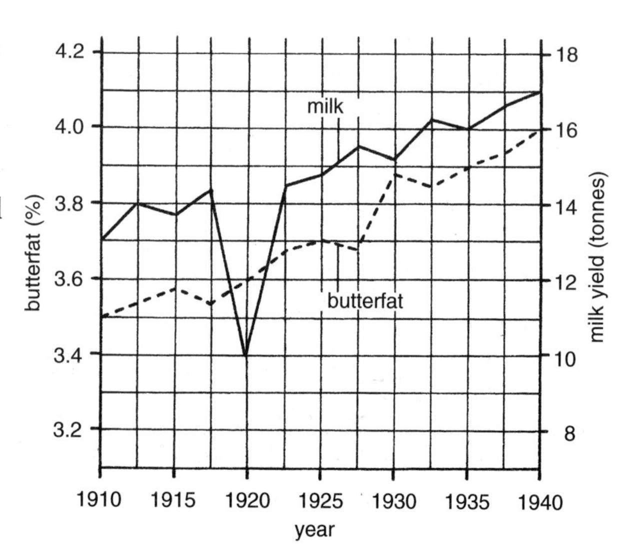

8 Between 1910 and 1940, the percentage butterfat content of milk increased by

A 0.4. **B** 0.5. **C** 5.0. **D** 12.5.

9 Between 1910 and 1935, milk yield (in tonnes) increased by

A 0.3. **B** 0.4. **C** 3.0. **D** 4.0.

Items 10 and 11 refer to the following list which gives some features of the processes of selective breeding and genetic engineering.

1 The organism's genotype becomes altered.
2 The new variety can make a substance previously only made by a different species.
3 The process involves working with many generations of the organism over a very long period of time.
4 The gene for a useful characteristic is transferred from one species to another.

10 Which of the following features refer to selective breeding?

A 1 and 2 **B** 1 and 3
C 2 and 3 **D** 1, 2 and 3

11 Which of the following features refer to genetic engineering?

A 1, 2 and 4 **B** 1, 3 and 4
C 2, 3 and 4 **D** 1, 2, 3 and 4

12 Which of the following diagrams correctly shows the action of a gene in a bacterial cell?

A

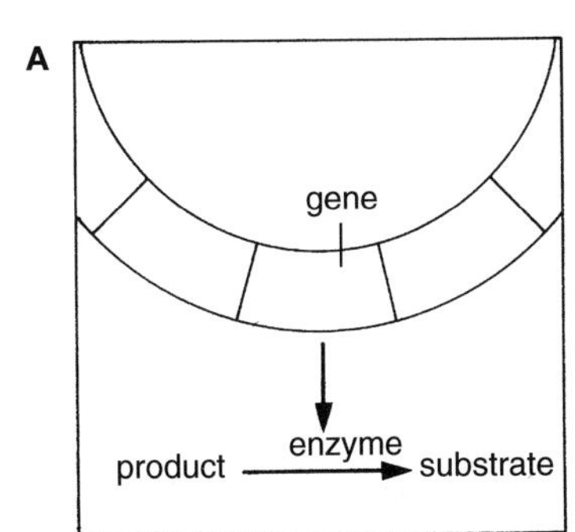

B

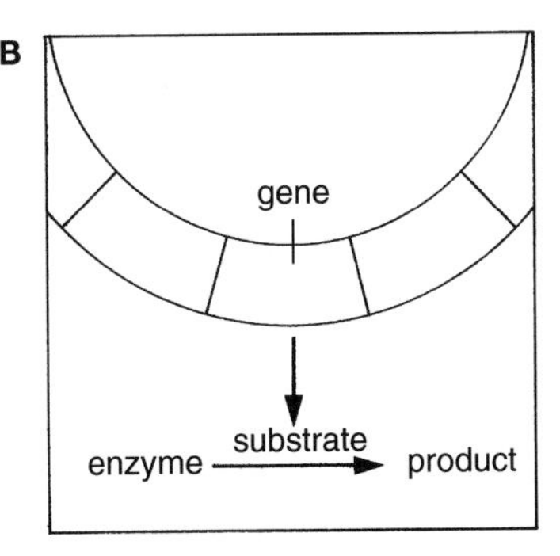

C

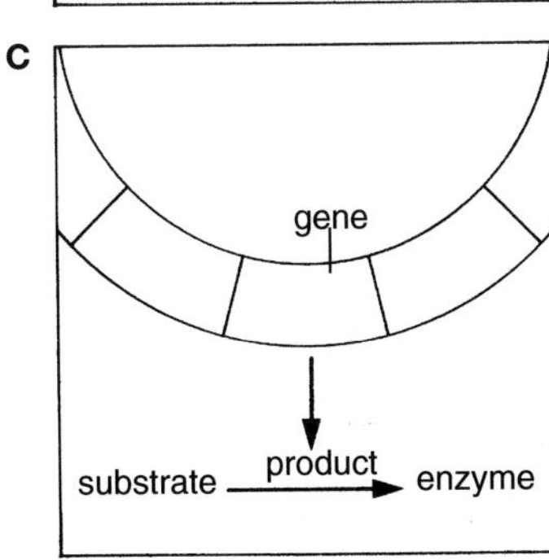

D

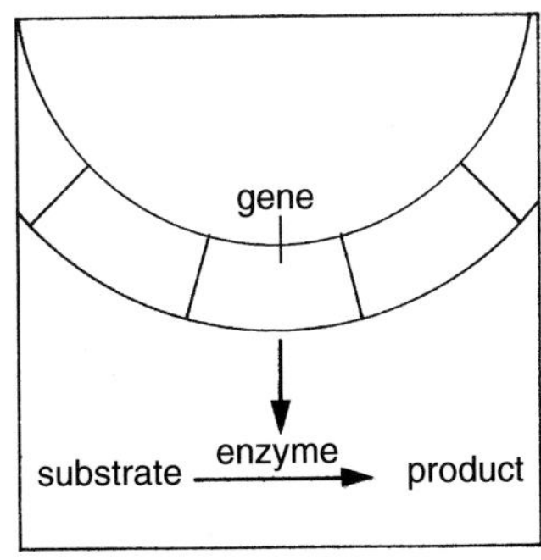

13 The substances listed below are products resulting from genetic engineering. Which substances BOTH have a medical application?

A Factor VIII and antifreeze chemical
B antifreeze chemical and rennin
C rennin and insulin
D insulin and Factor VIII

14 The first of the accompanying diagrams shows two preparatory stages carried out during the process of genetic engineering.

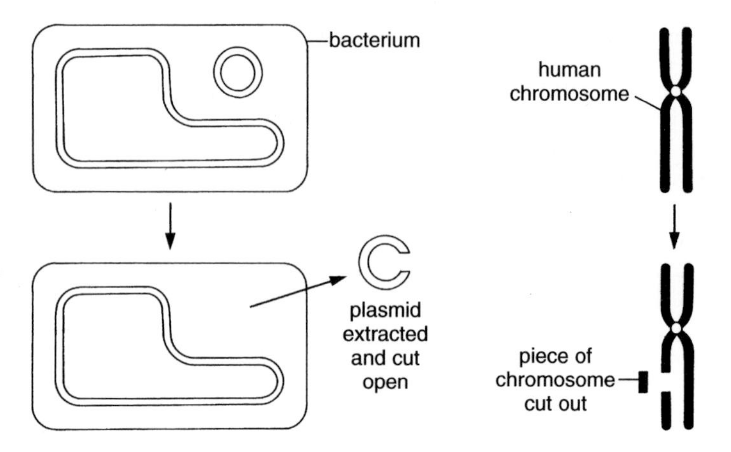

Which part of the second diagram shows a later stage in the process?

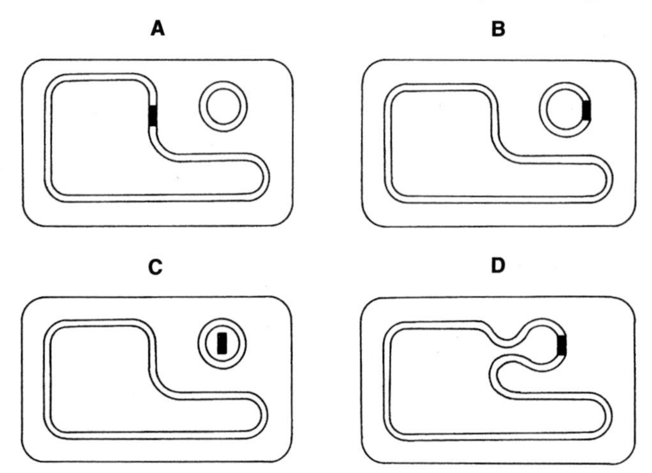

15 An example of a transgenic multicellular organism is a

A bacterium in the soil that invades wounded plant tissue.
B yeast cell engineered to make lager with a high alcohol content.
C person who would suffer diabetes mellitus without a supply of insulin.
D crop plant into which a useful gene has been successfully inserted.

11 Mammalian nutrition

Test 1

1 Carbohydrates, proteins and fats all contain the chemical elements

A nitrogen, carbon and hydrogen.
B oxygen, nitrogen and carbon.
C hydrogen, oxygen and nitrogen.
D carbon, hydrogen and oxygen.

2 Which of the following diagrams represents part of a molecule of cellulose?

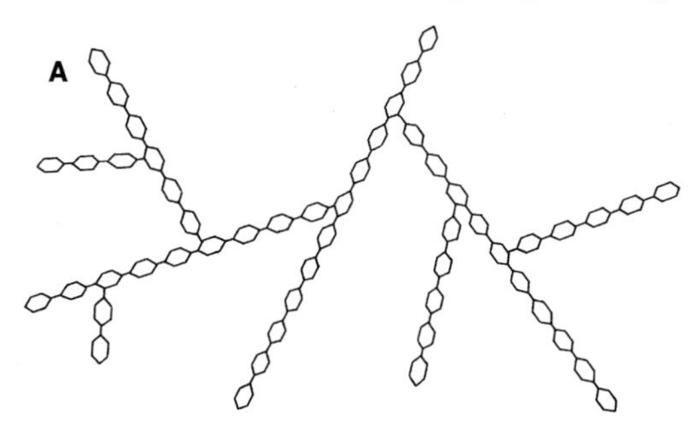

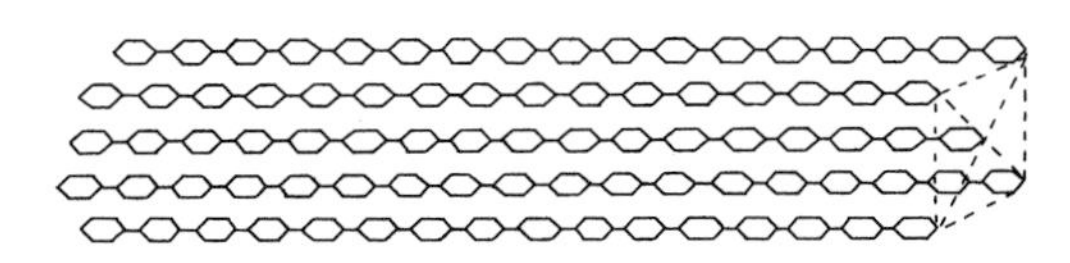

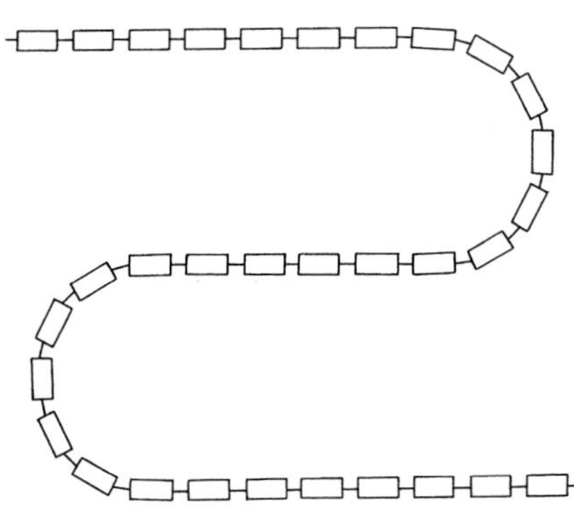

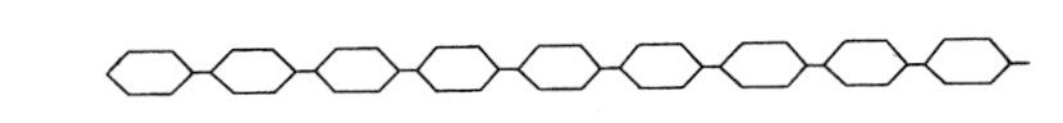

KEY

⬡ = glucose ▭ = amino acid

3 Which of the following is NOT a carbohydrate?

A starch **B** cellulose
C glycerol **D** glycogen

Items 4 and 5 refer to the following possible answers.

A fats **B** proteins
C vitamins **D** carbohydrates

4 Which food group releases most energy per gram?

5 Which food group is NOT a source of energy to the human body?

Items 6, 7 and 8 refer to the following possible answers.

A scurvy **B** rickets
C beri-beri **D** night blindness

6 Which of these conditions results when the human body suffers a deficiency of vitamin D?

7 Which condition is indicated by symptoms which include bleeding gums and poor healing of wounds?

8 Which condition is characterised by soft bones?

9 The recommended daily allowance (RDA) of vitamin B_1 for a young adult female is 1.2 mg. If 100 g of corn flakes contains 0.6 mg of vitamin B_1 then her RDA could be supplied by consuming a mass (in grams) of corn flakes of

A 2. **B** 50. **C** 72. **D** 200.

10 The accompanying table shows the calcium content (per 100 g) of several foodstuffs.

The minimum recommended intake of calcium for a 16-year-old is 600 mg. This could be achieved by eating

A 100 g almonds, 200 g oranges and 200 g chocolate biscuits.
B 100 g cottage cheese, 200 g ice cream and 400 g blackcurrants.
C 100 g milk chocolate, 200 g herring and 100 g yoghurt.
D 50 g sardines, 250 g white bread and 100 g milk.

Foodstuff	Calcium content (mg/100 g)
almonds	250
blackcurrants	60
chocolate biscuits	130
cottage cheese	80
herring	100
ice cream	140
milk	120
milk chocolate	250
oranges	40
sardines	400
white bread	100
yoghurt	140

Items 11 and 12 refer to the accompanying diagram of a model of part of the human alimentary canal.

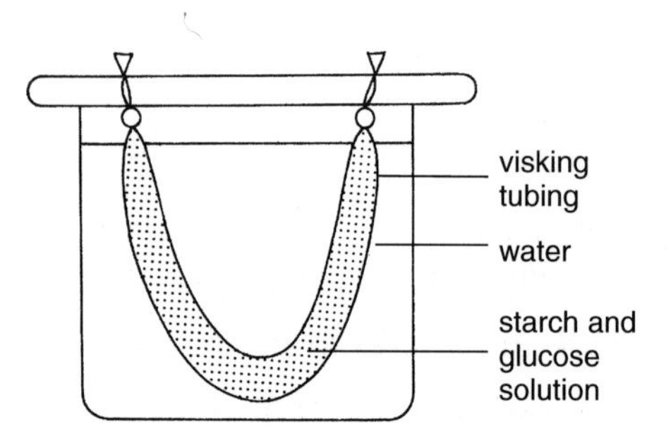

11 The water outside the visking tubing represents the

A water present in saliva.
B water surrounding the villi.
C blood supplying the intestine.
D blood supplying the oesophagus.

12 After 30 minutes, the water would contain

A glucose only.
B starch only.
C both glucose and starch.
D neither glucose nor starch.

13 Enzymes in the human alimentary canal are required to

A synthesise essential amino acids.
B absorb the end products of digestion.
C convert excess food to waste products.
D break down large molecules to soluble end products.

14 Which of the following diagrams correctly depicts the process of peristalsis?

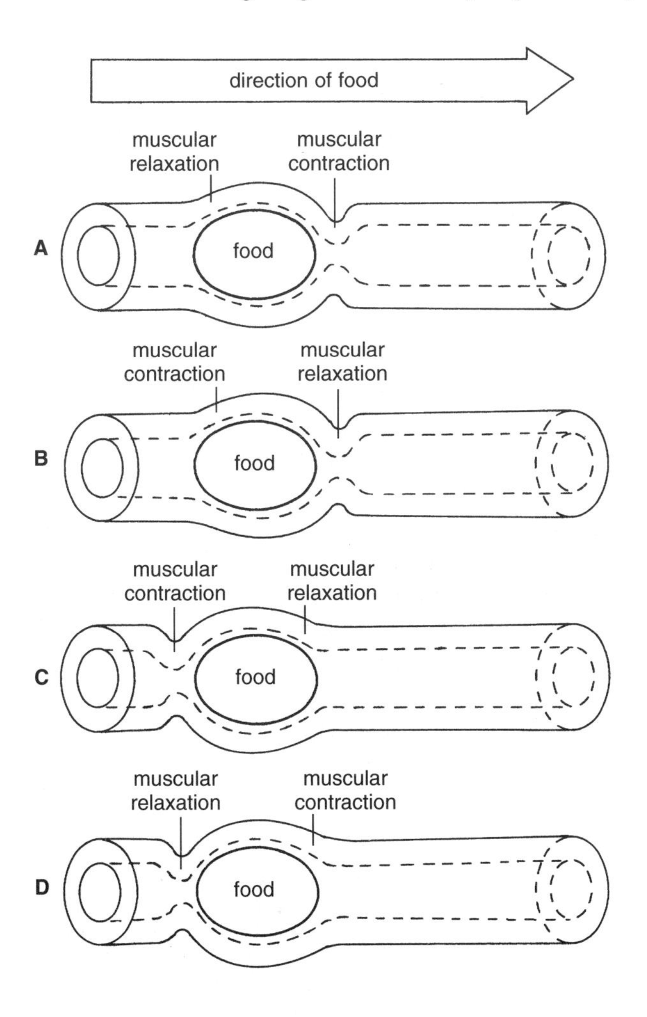

15 The accompanying diagram shows the positions of the sphincter valves at either end of the human stomach.

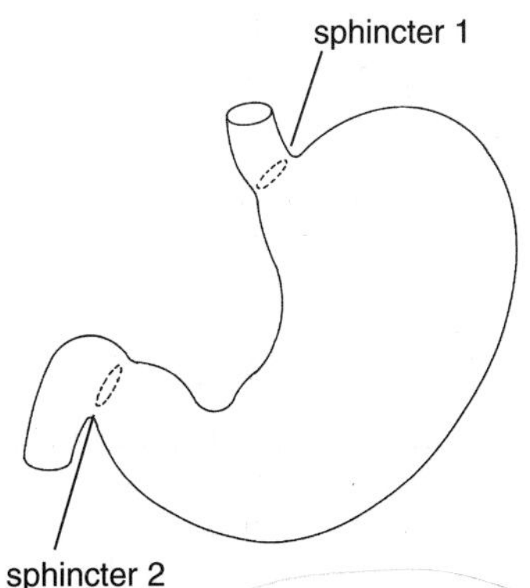

During the churning of food

A sphincter 1 is open and sphincter 2 is closed.
B sphincter 1 is closed and sphincter 2 is open.
C sphincters 1 and 2 are open.
D sphincters 1 and 2 are closed.

16 The accompanying diagram shows a gastric gland in the wall of the stomach. Which line in the table correctly matches the three cells with their secretions?

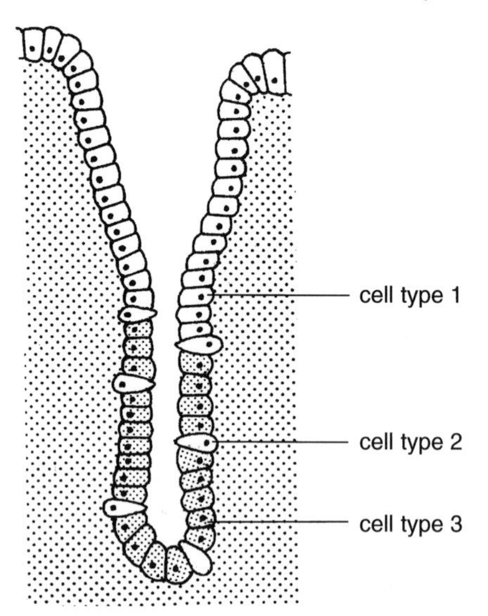

	Substance secreted		
	Enzyme	**Mucus**	**Acid**
A	2	1	3
B	3	1	2
C	1	3	2
D	3	2	1

Items 17, 18, 19 and 20 refer to the following graph of data obtained during an experiment where a person's stomach contents were sampled before, during, and at regular intervals after, a meal was eaten.

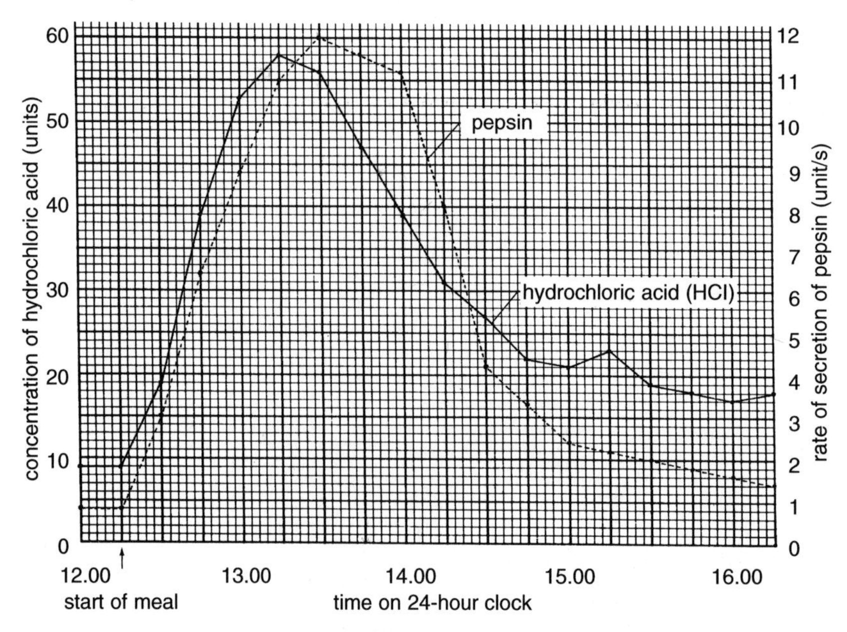

17 The time at which 56 units of hydrochloric acid and 12 units per second of pepsin were recorded was

A 13.15. **B** 13.30. **C** 13.45. **D** 14.00.

18 The time at which 44 units of hydrochloric acid and 11.4 units per second of pepsin were recorded was

A 12.51. **B** 13.21. **C** 13.51. **D** 13.57.

19 The readings recorded at 14.15 for the two substances were

	HCl (units)	Pepsin (units/s)
A	6.2	40.0
B	16.5	4.4
C	22.0	3.3
D	31.0	8.0

20 The number of minutes taken by pepsin to drop from its maximum rate of secretion to 5 units per second was

A 19. **B** 55. **C** 57. **D** 59.

21 The experiment shown in the first of the accompanying diagrams was carried out by a group of students to investigate the effect of acid on the action of the enzyme pepsin.

Other groups were asked to design a control experiment. Some of their suggestions are shown in the second diagram. Which of the four designs is a valid control?

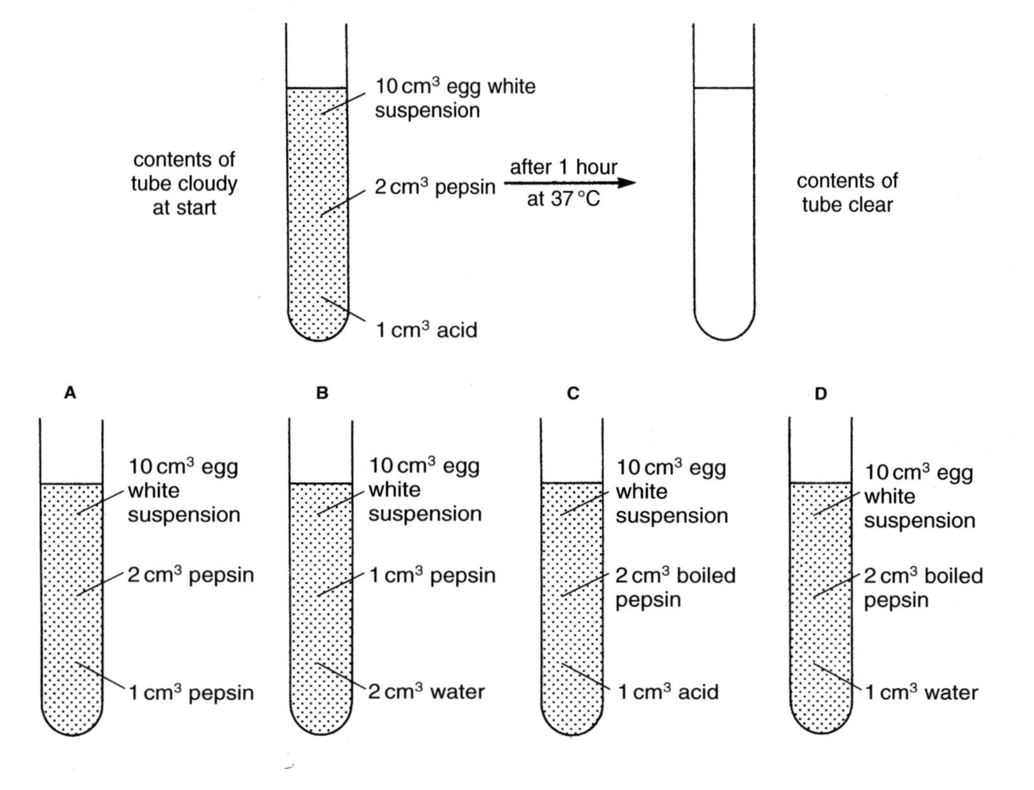

22 The accompanying diagram shows an experiment set up to investigate the effect of pH on the activity of the enzyme trypsin.

Which of the graphs that follow correctly represents the results?

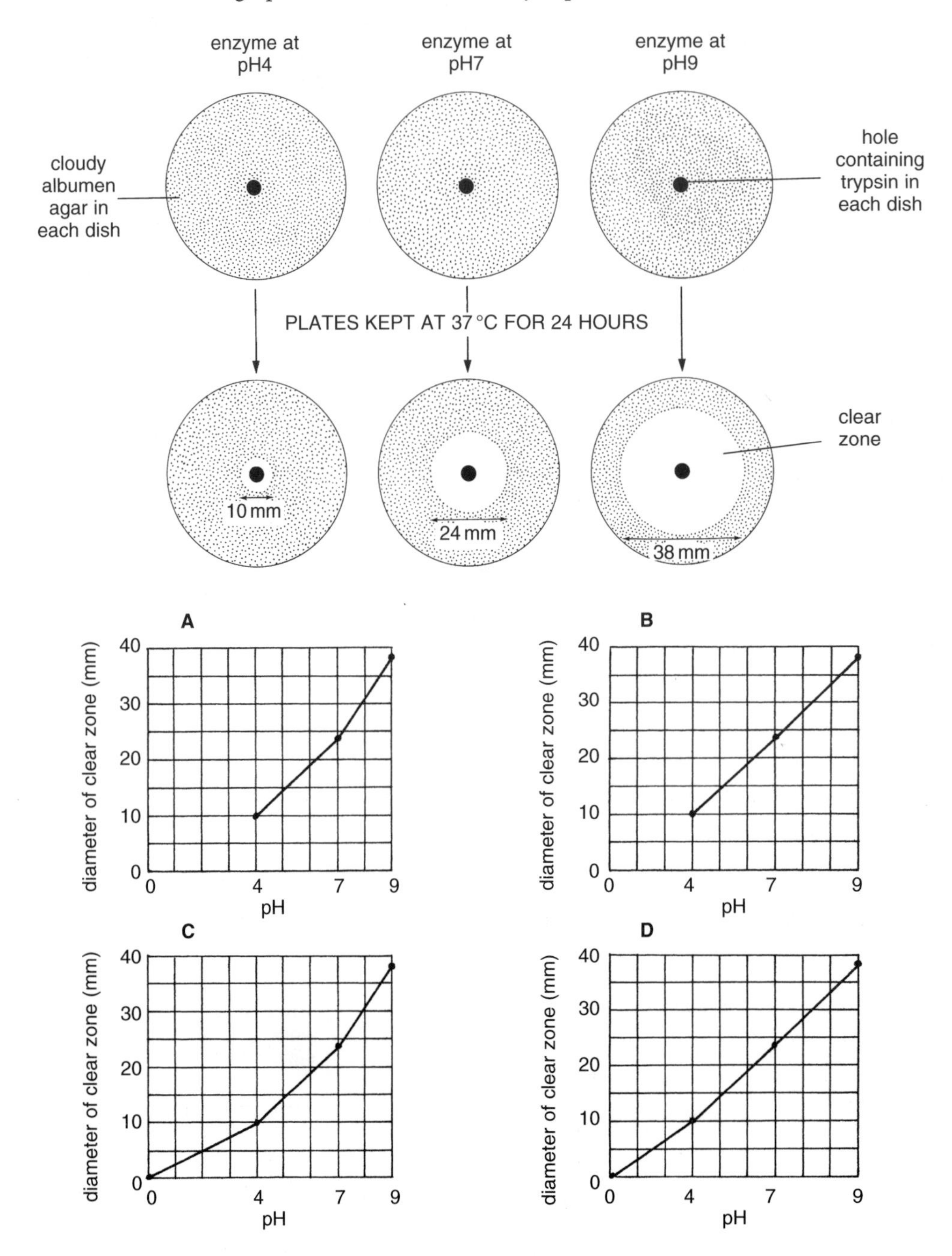

Items 23 and 24 refer to the following diagram of part of the small intestine.

23 Which structure brings about peristalsis?

24 Which structure is the site of absorption of digested food?

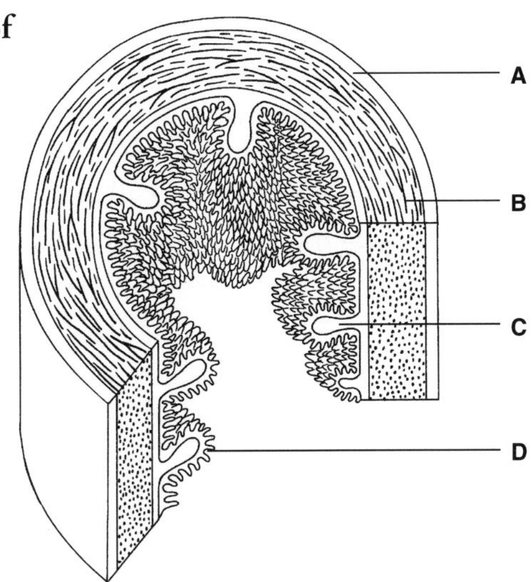

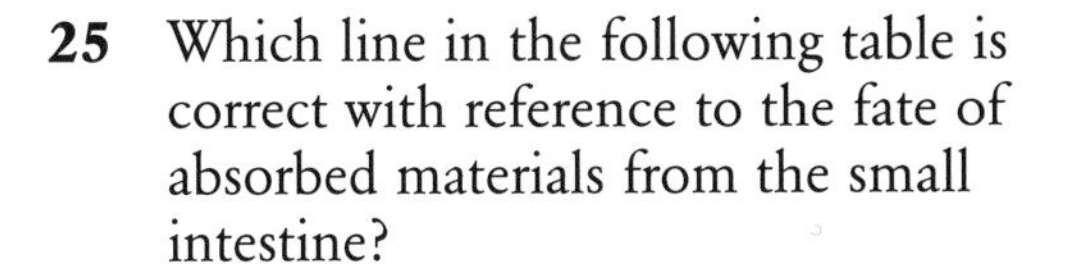

25 Which line in the following table is correct with reference to the fate of absorbed materials from the small intestine?

	Fate of excess glucose	Fate of excess amino acids
A	stored as glycogen in liver	converted to urea in liver
B	stored as glycogen in kidneys	converted to urea in kidneys
C	converted to urea in kidneys	stored as glycogen in kidneys
D	converted to urea in liver	stored as glycogen in liver

Test 2

1 Which of the following chemical elements is present in proteins but NOT in fats?

A carbon **B** oxygen
C hydrogen **D** nitrogen

2 The number of different types of amino acid commonly found to make up a protein is approximately

A 10. **B** 20. **C** 100. **D** 200.

3 Which of the following diagrams represents a molecule of fat?

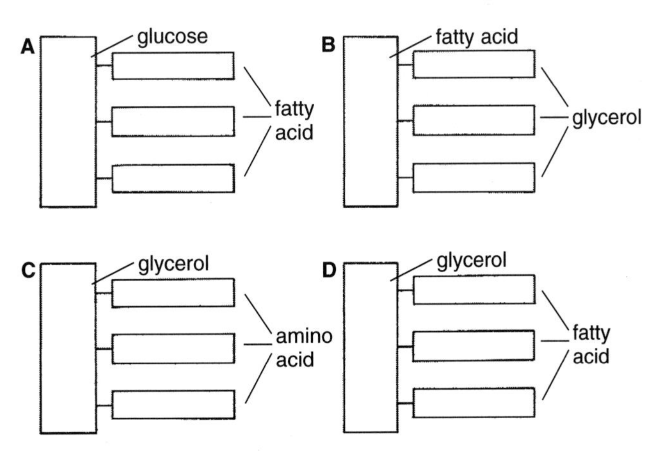

4 Which of the following statements is CORRECT?

A Starch is an insoluble carbohydrate stored in plant cells.
B Starch is a soluble carbohydrate stored in animal cells.
C Glycogen is a soluble carbohydrate stored in animal cells.
D Glycogen is an insoluble carbohydrate stored in plant cells.

Items 5, 6, 7, 8 and 9 refer to the information in the following table.

Foodstuff	Number of grams present per 100 g portion of food		
	Carbohydrate	Protein	Fat
bacon	0.0	11.0	48.0
brussels sprouts	4.6	3.6	0.0
cabbage	5.8	1.5	0.0
carrots	5.4	0.7	0.0
cheese	0.0	25.4	34.5
chocolate	54.5	8.7	37.6
coconut	6.4	6.6	62.3
eggs	0.0	11.9	12.3
fish fingers	20.7	13.4	6.8
liver (fried)	4.0	30.0	16.0
macaroni	84.0	9.9	1.0
milk	4.2	2.8	17.8
oranges	8.5	0.7	0.0
trifle	26.5	3.1	7.1

5 The number of foods that contain more carbohydrate than carrots is

A 6. **B** 7. **C** 10. **D** 12.

6 The number of foods that contain less protein than trifle is

A 4. **B** 6. **C** 8. **D** 11.

7 The number of times by which the fat content of coconut is greater than that of milk is

A 0.29. **B** 3.50. **C** 11.09. **D** 44.50.

8 A kilogram of chocolate would contain

	Fat (g)	Carbohydrate (g)
A	37.6	54.5
B	54.5	37.6
C	376.0	545.0
D	545.0	376.0

9 Which of the following stacked bar graphs represents fried liver?

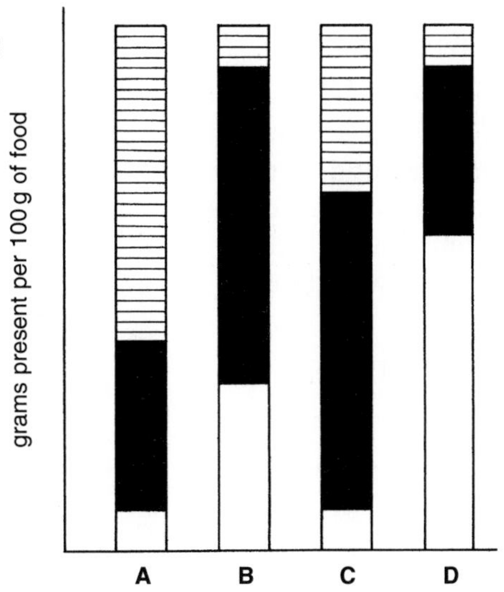

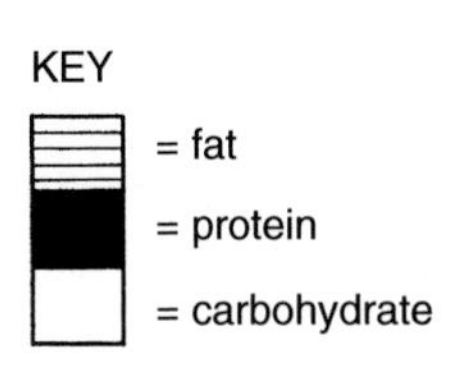

10 It is INCORRECT to say that vitamins are

A continuously reused by living cells.
B required to promote biochemical reactions.
C only required in small quantities.
D needed to provide the body with energy.

11 Which of the following foods is a rich source of vitamin A?

A yeast **B** oranges
C cod liver oil **D** carrots

12 The recommended daily allowance (RDA) of vitamin C for a pregnant woman is 80 mg. If this RDA is supplied exactly by consuming 125 g of strawberries then the mass in mg of vitamin C present in 100 g of strawberries is

A 6.4. **B** 15.6. **C** 64.0. **D** 100.0.

13 Which of the following mineral elements is needed for production of hormones by the thyroid gland?

A iron **B** iodine
C sodium **D** phosphorus

14 The experiment shown in the accompanying diagram is set up to investigate the effect of the enzyme salivary amylase on starch.

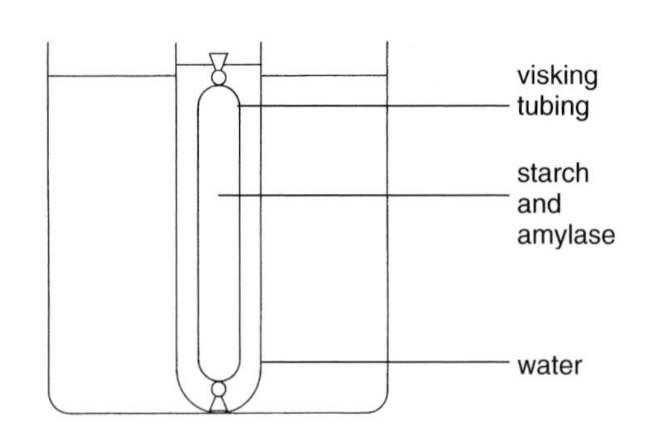

If samples of the water surrounding the visking tubing are tested at the start of the experiment and after one hour, which of the following sets of results would be obtained?

+ = food present – = food absent

A

	Starch test	Sugar test
At start	–	–
After one hour	–	+

B

	Starch test	Sugar test
At start	–	+
After one hour	–	+

C

	Starch test	Sugar test
At start	–	–
After one hour	+	–

D

	Starch test	Sugar test
At start	–	–
After one hour	+	+

Items 15 and 16 refer to the accompanying diagram.

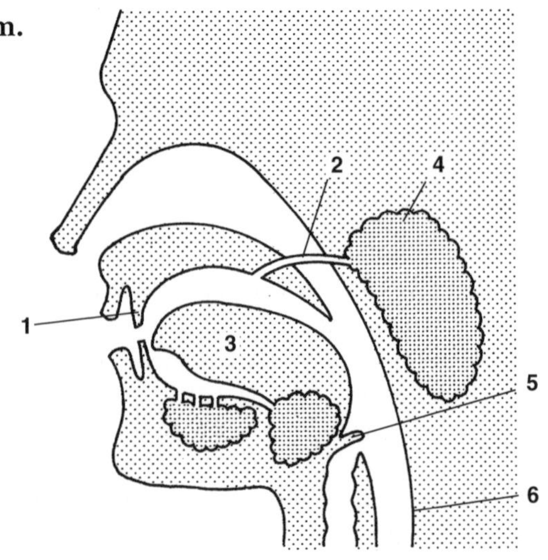

15 The structure responsible for the mechanical breakdown of food is numbered.

A 1. **B** 2. **C** 3. **D** 5.

16 Mucus is manufactured by structure

A 2. **B** 3. **C** 4. **D** 6.

17 Which line in the following table correctly refers to peristalsis?

	State of circular muscle	Diameter of alimentary canal round food bolus
A	contracted in front of food bolus	wider
B	relaxed behind food bolus	narrower
C	contracted behind food bolus	wider
D	relaxed in front of food bolus	narrower

18 The accompanying diagram shows a simplified version of the human stomach. Which of the diagrams that follow shows its appearance following contraction of the layer of longitudinal muscle in the stomach wall?

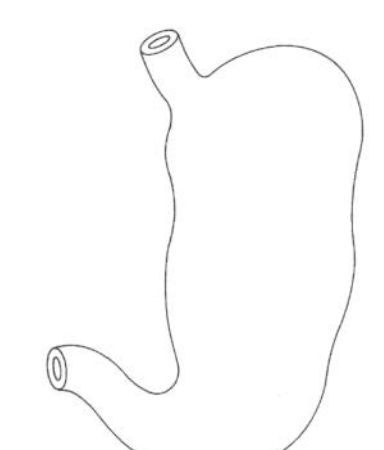

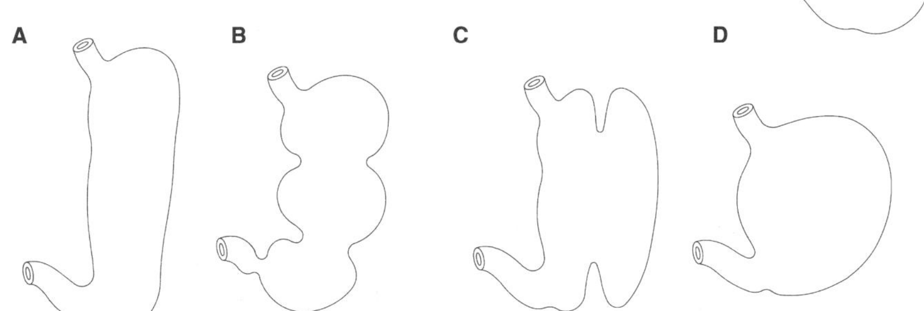

19 Which of the following statements about the stomach and its contents is INCORRECT?

A Acid is needed to convert the enzyme pepsin to pepsinogen.
B Mucus protects the stomach lining from damage by enzymes.
C Pepsin digests molecules of complex protein to simpler peptides.
D Movement of the stomach wall brings about churning of food.

20 The accompanying bar graph shows the results of an experiment where the contents of the stomachs of a large number of people were examined for acid content following the consumption of a 100 g portion of one of six foods.

It can be concluded from the data that the stomach secretes

A less acid when food rich in protein is consumed.
B more acid when fruit and vegetables are eaten.
C less acid when a diet rich in fat is consumed.
D more acid when food rich in protein is eaten.

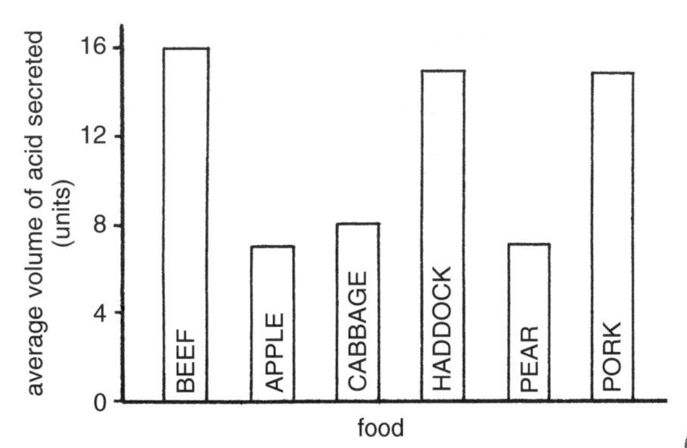

Items 21, 22 and 23 refer to the following information. The test tubes shown in the accompanying diagram were set up to investigate the effect of boiling and the effect of acid on the activity of pepsin.

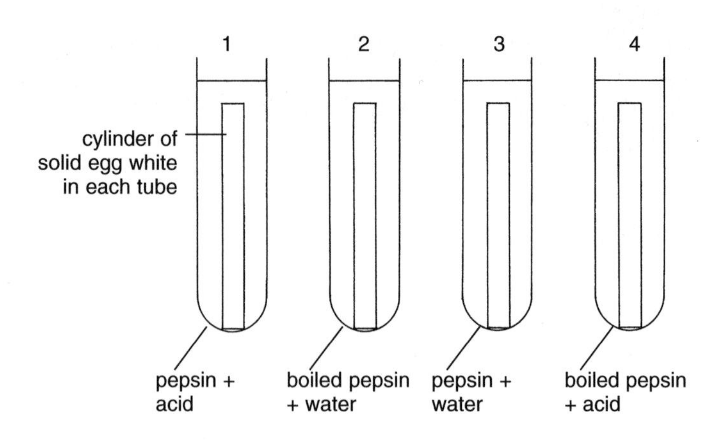

21 Which two tubes should be compared at the end of the experiment to draw a conclusion about the effect of acid on pepsin's activity?

A 1 and 2 **B** 1 and 3
C 1 and 4 **D** 2 and 4

22 Which two test tubes should be compared at the end of the experiment to draw a conclusion about the effect of temperature on pepsin's activity?

A 1 and 2 **B** 1 and 4
C 2 and 4 **D** 3 and 4

23 Which test tube could have been omitted without affecting the validity of the experiment?

A 1 **B** 2 **C** 3 **D** 4

Items 24 and 25 refer to the accompanying diagram of a close-up of the surface of the small intestine.

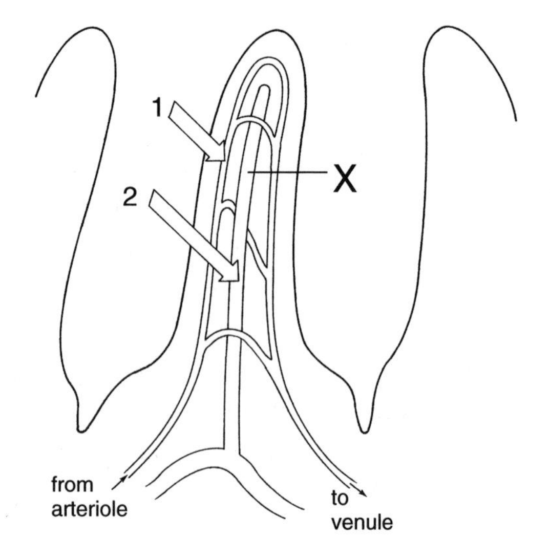

24 Structure X is called

A a lacteal. **B** an epithelium.
C a goblet cell. **D** a blood capillary.

25 The products of protein digestion enter by route

A 1 and are transported in the lymphatic system.
B 2 and are transported in the blood circulatory system.
C 1 and are transported in the blood circulatory system.
D 2 and are transported in the lymphatic system.

Control of internal environment

Test 1

1 Which of the following terms means the maintenance of the body's water balance at a normal level?

A excretion **B** filtration
C reabsorption **D** osmoregulation

Items 2 and 3 refer to the data in the following tables of the daily water balance of a 16-year-old during the summer holidays.

Table 1

Source of water gain	Volume gained (cm^3)
food	950
drink	
metabolic reactions	350

Table 2

Source of water loss	Volume lost (cm^3)
sweat	500
urine	1500
exhaled air	400
faeces	100

2 The volume of water (in cm^3) gained by this person in drink was

A 600. **B** 700. **C** 1200. **D** 1300.

3 Which of the following pie charts correctly represents the data in Table 2?

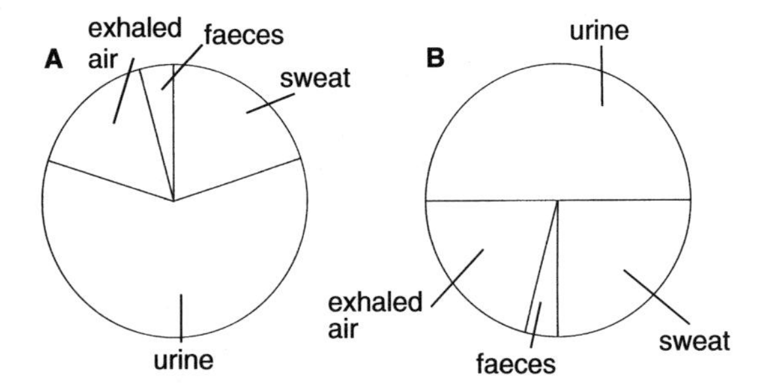

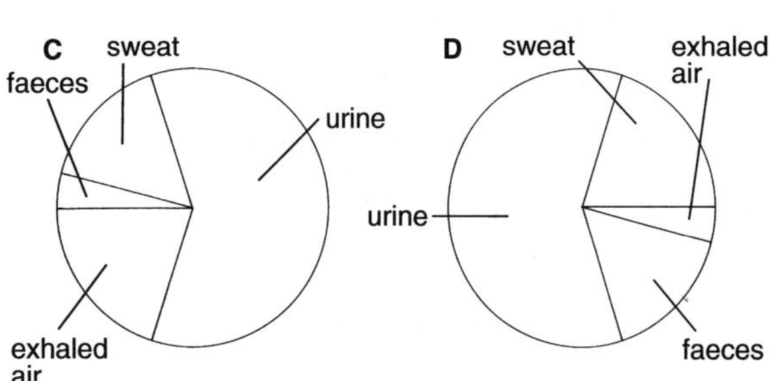

4 The accompanying diagram shows part of a mammal's excretory system.

Which structure removes blood from the kidney?

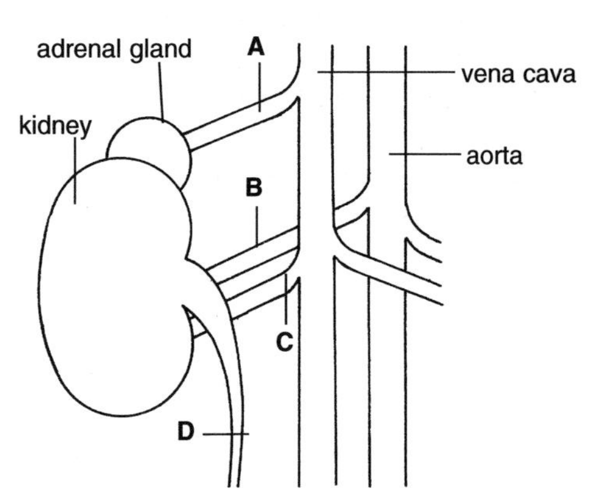

5 Urease is an enzyme which breaks down urea. This process results in the release of a gas which turns red litmus paper blue.

The accompanying diagram shows an experiment set up to investigate the chemical composition of three liquids. In which tube(s) would the litmus paper turn blue?

A 3 only **B** 1 and 3
C 2 and 3 **D** 1 and 2

1 2 3
moist litmus paper (red at start in each tube)
tablet of urease enzyme in each tube
solution equivalent to urine
solution equivalent to plasma from renal artery
solution equivalent to plasma from renal vein

Items 6, 7, 8, 9 and 10 refer to the accompanying diagram of a nephron.

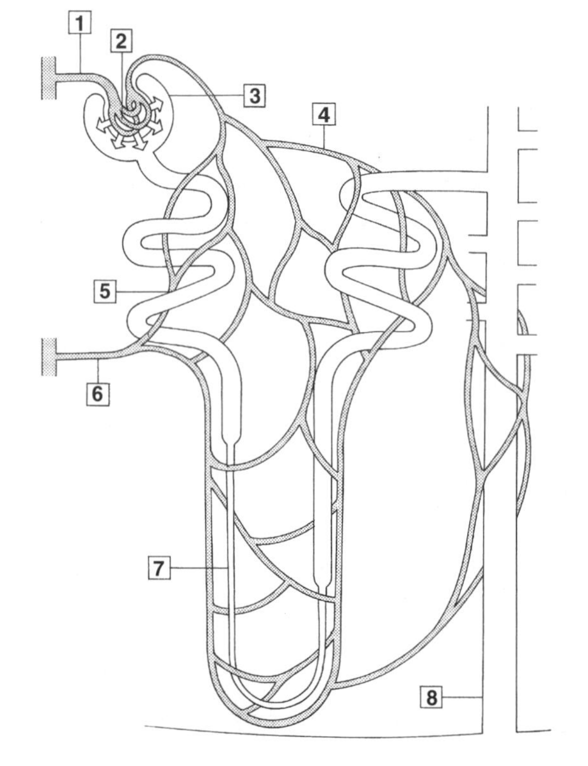

6 The glomerulus is labelled

A 2. **B** 3. **C** 5. **D** 7.

7 The Bowman's capsule is labelled

A 2. **B** 3. **C** 5. **D** 7.

8 The collecting duct is labelled

A 3. **B** 6. **C** 7. **D** 8.

9 The structure which transports urea to the nephron is labelled

A 1. **B** 4. **C** 6. **D** 8.

10 Glucose is reabsorbed from the filtrate at point

A 2. **B** 3. **C** 5. **D** 8.

11 Which of the following factors helps to maintain high blood pressure in a glomerulus?

A The vessel leaving a glomerulus is narrower than the one entering it.
B Plasma proteins in the bloodstream tend to force small molecules out of the blood.
C Filtrate present in the Bowman's capsule tends to draw further filtrate from the bloodstream by osmosis.
D The blood vessel supplying a glomerulus contains blood arriving from the renal vein.

Items 12, 13 and 14 refer to the accompanying table which compares the composition of three liquids present in the body of a student volunteer.

Substance present in liquid	Percentage of substance in blood plasma	Percentage of substance in glomerular filtrate	Percentage of substance in urine
W	0.03	0.03	2.00
X	0.10	0.10	0.00
Y	8.00	0.00	0.00
Z	0.42	0.42	1.05

12 Which substance was filtered into the kidney tubules and then completely reabsorbed back into the bloodstream?

A W **B** X **C** Y **D** Z

13 Which substance was NOT filtered out of the bloodstream?

A W **B** X **C** Y **D** Z

14 Compared with its concentration in blood plasma and glomerular filtrate, the concentration of substance Z in urine has increased by a factor of

A 0.40. **B** 0.44. **C** 0.63. **D** 2.50.

15 A woman produced 150 litres of glomerular filtrate and 1.5 litres of urine during a 24-hour period. The percentage of water reabsorbed into her bloodstream was

A 1. **B** 10. **C** 90. **D** 99.

16 Which of the following correctly describes an osmoregulatory problem?

A A saltwater bony fish gains water from its hypotonic surroundings.
B A saltwater bony fish loses water to its hypertonic surroundings.
C A freshwater bony fish gains water from its hypertonic surroundings.
D A freshwater bony fish loses water to its hypotonic surroundings.

17 Which of the osmoregulatory mechanisms shown in the accompanying diagram is employed by a freshwater bony fish?

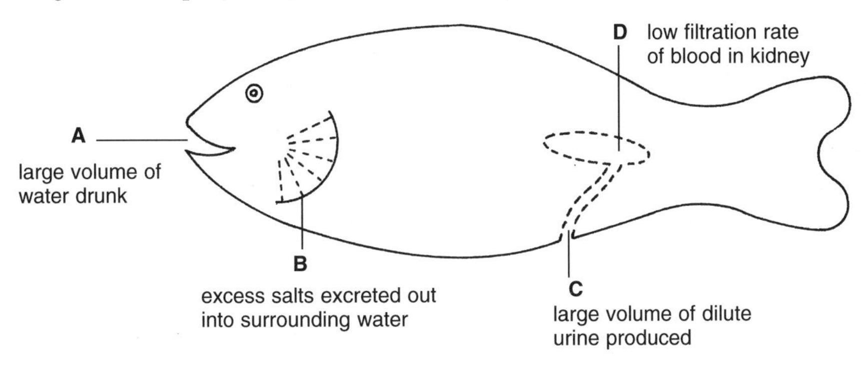

18 The gills of a saltwater bony fish

A lose water by osmosis and absorb salts.
B lose water by osmosis and excrete salts.
C gain water by osmosis and absorb salts.
D gain water by osmosis and excrete salts.

19 The structures shown in the accompanying diagram are present in the kidneys of a certain animal.

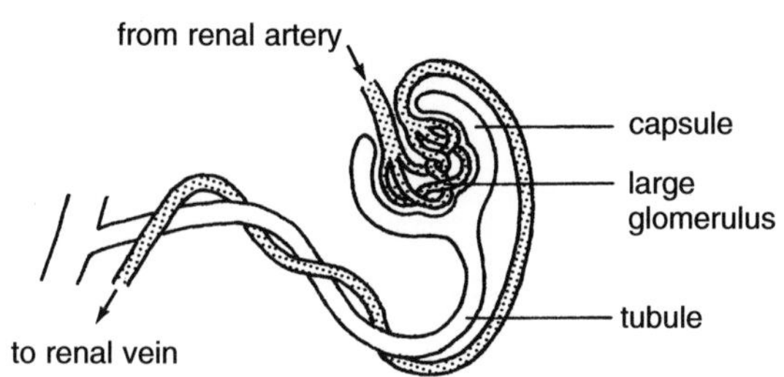

This could be a

A desert rat.
B human being.
C freshwater bony fish.
D saltwater bony fish.

Items 20 and 21 refer to the possible answers given in the following table.

	Relative quantity of water drunk	Relative quantity of urine produced
A	none	little
B	none	much
C	much	much
D	much	little

20 Which animal is a cod from the North Sea?

21 Which animal is a desert rat?

22 Negative feedback control involves the following four stages.

1 Effectors bring about corrective responses.
2 A receptor detects a change in the internal environment.
3 Variation from the norm is counteracted.
4 Nerve or hormonal messages are sent to effectors.

The order in which these occur is

A 2, 4, 1, 3. **B** 2, 4, 3, 1.
C 4, 2, 1, 3. **D** 4, 2, 3, 1.

Items 23 and 24 refer to the accompanying diagram of osmoregulation in the human body.

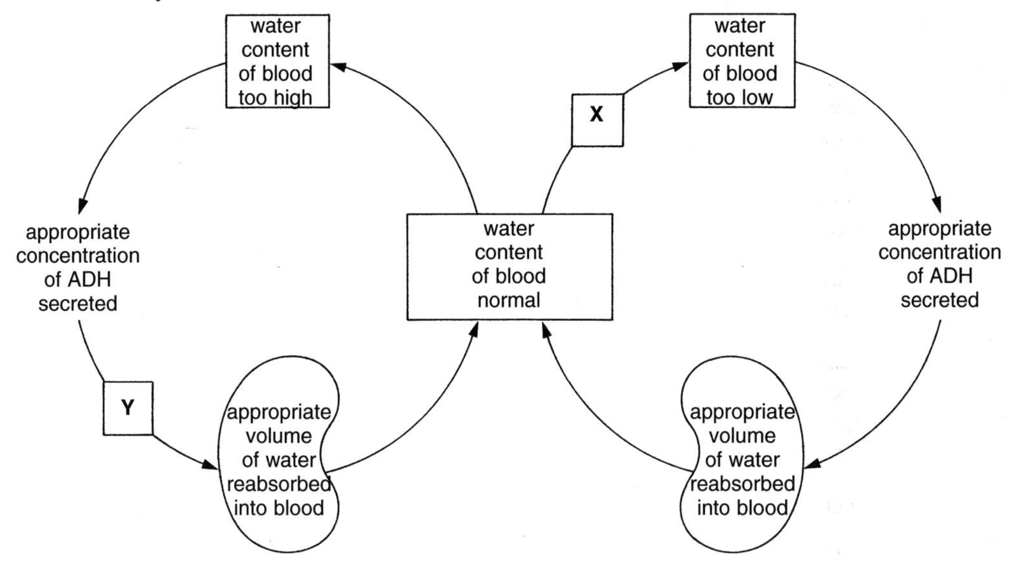

23 Which of the following took place at point X to bring about the change indicated?

A The person passed urine.
B The person ate salty food.
C The person consumed a fizzy drink.
D The person's rate of sweating decreased.

24 Which of the following BOTH occurred at point Y?

	Concentration of ADH in blood	Rate of reabsorption of water into blood
A	increased	increased
B	increased	decreased
C	decreased	increased
D	decreased	decreased

25 When a decrease in water concentration of blood occurs, which of the following series of events brings about osmoregulation?

	ADH production	Permeability of kidney tubules	Volume of urine produced
A	↑	↑	↓
B	↑	↓	↓
C	↓	↑	↓
D	↑	↑	↑

↑ = increases ↓ = decreases

Test 2

1 The accompanying bar graph represents selected parts of the data given in the following tables of a student's average daily water balance.

Source of water loss	Volume lost (cm^3)
urine	1700
exhaled air	475
sweat	525
faeces	100

Source of water gain	Volume gained (cm^3)
food	625
drink	1750
metabolic reactions	425

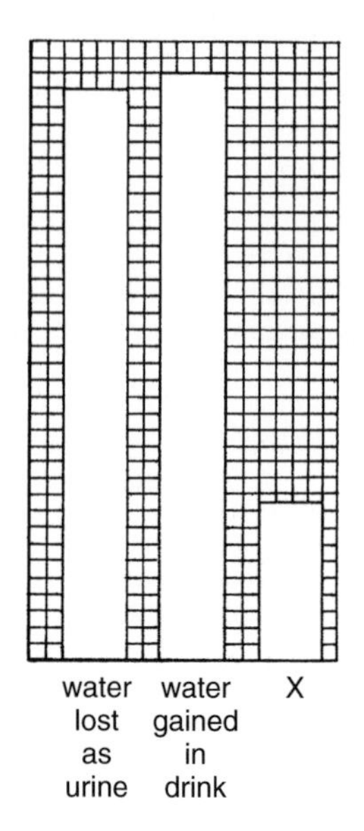

Bar X in the graph represents

A water lost in sweat.
B water gained in food.
C water lost in exhaled air.
D water gained in metabolic reactions.

2 The main organ for regulating a mammal's water content is the

A skin. **B** lung.
C colon. **D** kidney.

3 Which line in the following table is CORRECT?

	Food from which urea is produced	Site of production of urea	Site of removal of urea from bloodstream
A	protein	kidney	liver
B	fat	liver	kidney
C	protein	liver	kidney
D	fat	kidney	liver

Questions 4, 5, 6, 7, 8 and 9 refer to the accompanying diagram of a mammal's excretory system.

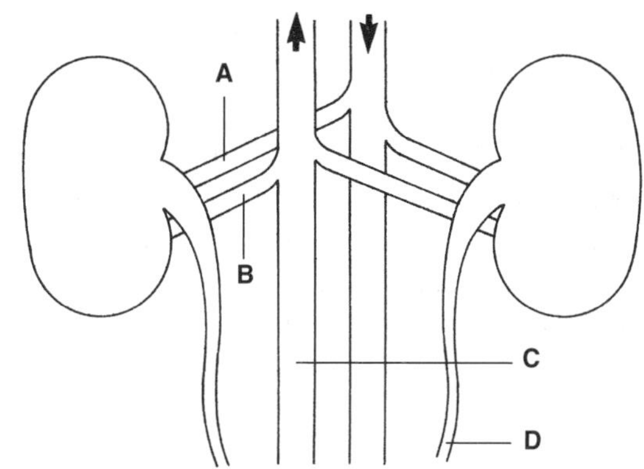

4 Which letter points to the ureter?

5 Which letter indicates the vena cava?

6 Which letter points to the renal artery?

7 Which letter indicates the structure responsible for transporting blood to a kidney?

8 Which letter points to the renal vein?

9 Which letter indicates the tube that carries urine to the bladder?

10 The following table compares the composition of blood plasma in a renal artery with that in a renal vein. Which line in the table is INCORRECT?

	Substance	Relative concentration of substance in renal artery	Relative concentration of substance in renal vein
A	oxygen	higher	lower
B	urea	lower	higher
C	CO_2	lower	higher
D	salts	higher	lower

Questions 11 and 12 refer to the following answers.

A glomerulus **B** kidney tubule
C collecting duct **D** capillary network

11 Which term means a communal tube shared by several nephrons?

12 Which term means a knot of blood vessels surrounded by a Bowman's capsule?

Questions 13, 14 and 15 refer to the accompanying diagram of a kidney nephron and its blood supply.

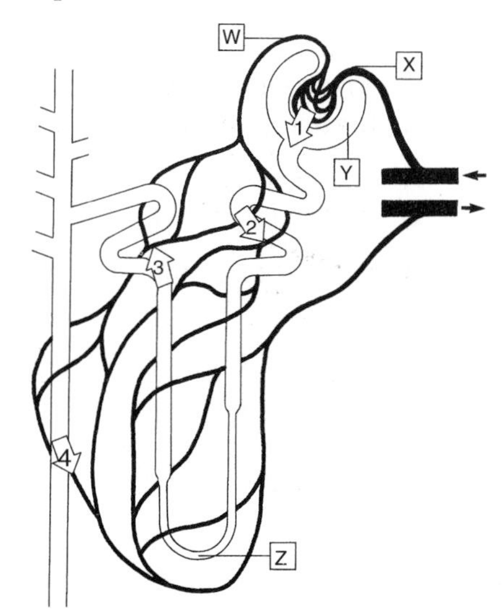

13 Filtration of blood takes place at arrow

A 1. **B** 2. **C** 3. **D** 4.

14 Reabsorption of useful substances occurs at arrow(s)

A 1 only. **B** 2 and 3 only.
C 2, 3 and 4 only. **D** 1, 2, 3 and 4.

15 Blood at the highest pressure is found at position

A W. **B** X. **C** Y. **D** Z.

16 Which line in the following table refers correctly to the urea concentration present in three liquids from a healthy human adult?

	Concentration of urea (g/100 cm³)		
	Blood plasma	**Glomerular filtrate**	**Final urine**
A	0.03	0.03	2.1
B	0.03	2.1	2.1
C	2.1	0.03	0.03
D	2.1	2.1	0.03

17 An adult woman produced a total of 144 litres of glomerular filtrate in 24 hours. Her GFR (volume of glomerular filtrate in millilitres produced per minute by both kidneys) was

A 0.1. **B** 1.0. **C** 10.0. **D** 100.0.

Questions 18, 19 and 20 refer to the information in the following table.

Species of animal	**Group of animal kingdom**	**Type of water in natural environment**
Paramecium	invertebrate	fresh
jellyfish	invertebrate	salt
stickleback	vertebrate	fresh
haddock	vertebrate	salt

18 Which animal possesses body contents which are isotonic to its natural environment?

A *Paramecium* **B** jellyfish
C stickleback **D** haddock

19 Which animal possesses body contents which are hypotonic to its natural environment?

A *Paramecium* **B** jellyfish
C stickleback **D** haddock

20 Which pair of animals possess body contents that are hypertonic to their natural environment?

A *Paramecium* and jellyfish
B *Paramecium* and stickleback
C jellyfish and haddock
D stickleback and haddock

21 Which of the osmoregulatory mechanisms shown in the accompanying diagram is employed by a saltwater fish?

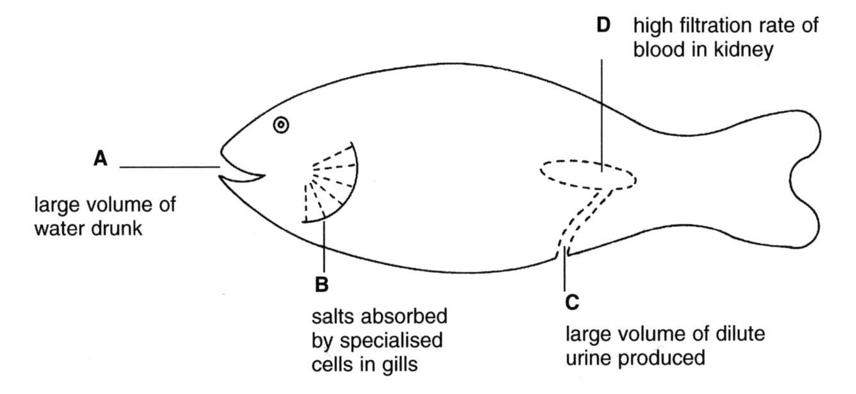

22 The water concentration of a freshwater fish's body is maintained at the optimum level by

A a large volume of the surrounding water being drunk.
B mineral salts being excreted by specialised gill cells.
C a small volume of concentrated urine being produced.
D blood passing through the kidneys being filtered at a high rate.

23 The accompanying diagram illustrates the principle of negative feedback control. The words that should have been inserted in boxes **X** and **Y** are

	X	Y
A	receptor	receptor
B	receptor	effector
C	effector	receptor
D	effector	effector

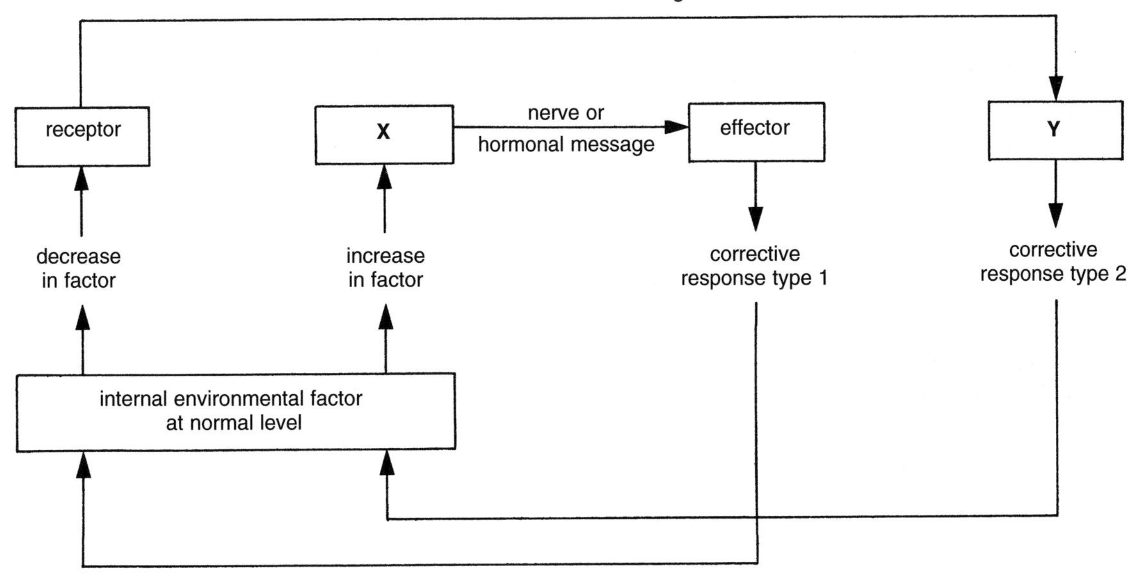

24 A man who is not thirsty drinks a litre of water. Which line in the following table correctly summarises the events that will result from this behaviour?

	ADH production	Water reabsorption	Urine output
A	↑	↑	↓
B	↑	↓	↑
C	↓	↓	↑
D	↓	↑	↓

↑ = increases ↓ = decreases

25 Which of the following is correct?

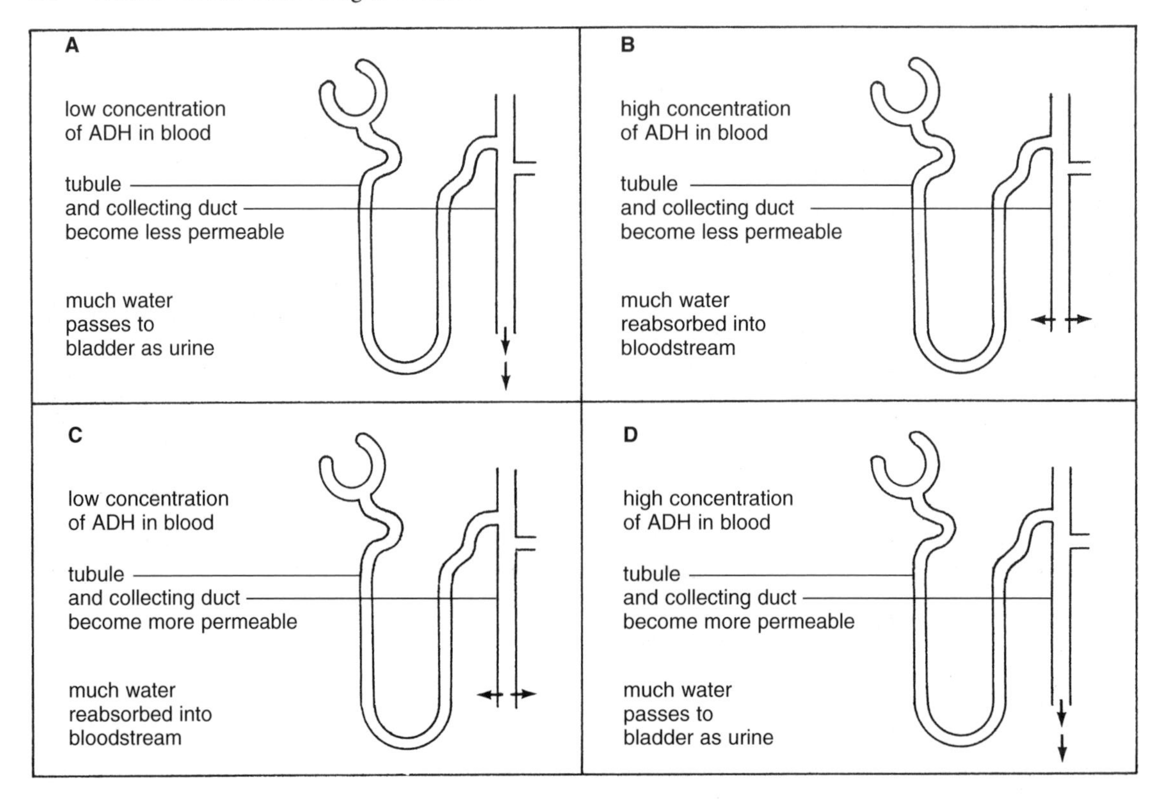

13 Circulation and gas exchange

Test 1

Items 1, 2, 3 and 4 refer to the accompanying diagram of a mammalian heart.

1 Chamber P is called the

A left atrium and it receives blood from all parts of the body.
B right atrium and it receives blood from the lungs.
C left atrium and it receives blood from the lungs.
D right atrium and it receives blood from all parts of the body.

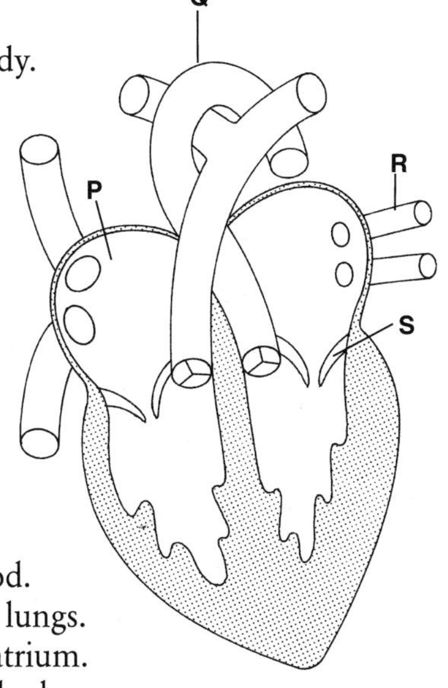

2 Vessel Q carries

A oxygenated blood from the heart.
B deoxygenated blood to the lungs.
C oxygenated blood from the lungs.
D deoxygenated blood to the heart.

3 Vessel R is the

A aorta. **B** vena cava.
C pulmonary vein. **D** pulmonary artery.

4 The function of structure S is to

A prevent oxygenated blood mixing with deoxygenated blood.
B direct the flow of deoxygenated blood returning from the lungs.
C prevent oxygenated blood passing from a ventricle to an atrium.
D direct the flow of deoxygenated blood from the heart to the lungs.

5 Which blood vessels BOTH possess semi-lunar (SL) valves?

A vena cava and aorta
B aorta and pulmonary artery
C pulmonary artery and pulmonary vein.
D pulmonary vein and vena cava.

6 Which line in the following table is CORRECT?

	Relative thickness of wall		
	Right ventricle (RV)	**Left ventricle (LV)**	**Reason for difference**
A	thicker	thinner	RV pumps blood all round body
B	thinner	thicker	LV pumps blood all round body
C	thicker	thinner	LV pumps blood all round body
D	thinner	thicker	RV pumps blood all round body

7 The coronary arteries supply

A oxygenated blood to the heart wall muscle.
B deoxygenated blood to the heart wall muscle.
C oxygenated blood to the ventricle chambers.
D deoxygenated blood to the ventricle chambers.

8 The accompanying table shows some of the findings from an international survey carried out during the 1990s on death rate from heart disease of people aged between 45 and 69 years.

Country	Death per 100 000 population	
	Women	Men
Japan	10	51
France	35	115
Germany	64	246
Australia	68	340
Scotland	140	510

Which of the following statements is CORRECT?

A Compared with Australian men, the death rate amongst Scottish men is 1.5 times lower.
B Compared with German women, the death rate amongst Scottish women is 3 times higher.
C Compared with Scottish women, the death rate amongst French women is 4 times lower.
D Compared with Scottish men, the death rate amongst Japanese men is 10 times higher.

Questions 9 and 10 refer to the accompanying graph which records the pulse rate of a woman taken at one-minute intervals before, during and after a period of vigorous exercise.

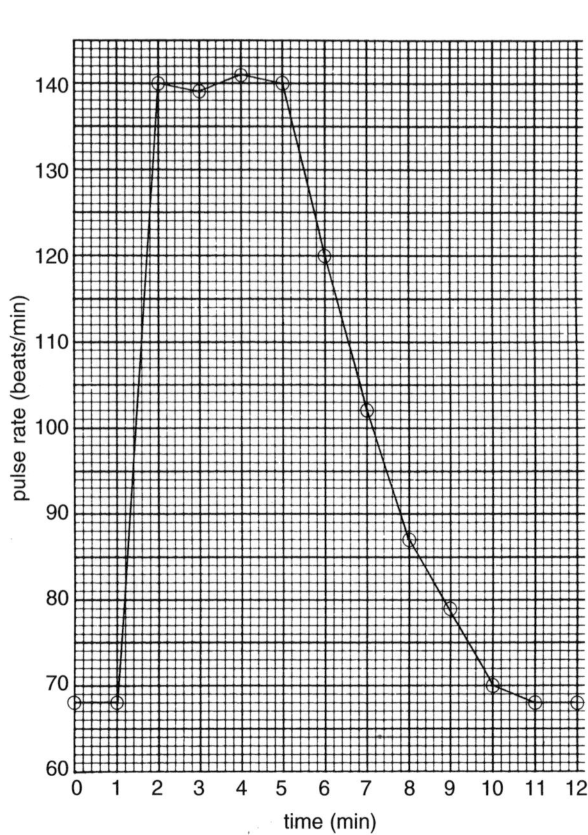

9 For how many minutes did the period of vigorous exercise last?

A 1 **B** 2 **C** 3 **D** 4

10 During which time interval in minutes did pulse rate decrease at the fastest rate?

A 5–6 **B** 6–7 **C** 7–8 **D** 8–9

Items 11, 12 and 13 refer to the following list of features of certain blood vessels.

1 dissolved food and oxygen diffuse through their walls
2 blood inside them is being transported back to the heart
3 blood carried by them is at low pressure
4 blood carried by them is at high pressure
5 their walls are only one cell thick
6 their walls are thick and elastic

11 Which features BOTH apply to the capillaries?

A 1 and 6 **B** 1 and 5
C 2 and 5 **D** 2 and 6

12 Which features BOTH apply to arteries?

A 1 and 2 **B** 3 and 6
C 1 and 4 **D** 4 and 6

13 Which features BOTH apply to veins?

A 2 and 3 **B** 2 and 4
C 4 and 5 **D** 3 and 6

14 The data in the accompanying table refer to blood vessels.

Blood vessel	Average external diameter (μm)	Average internal diameter (μm)
capillary	10	8
artery	3000	336
vein	3600	3000

The internal diameter of a vein is greater than that of a capillary by a factor of

A 42. **B** 360. **C** 375. **D** 450.

Questions 15 and 16 refer to the following possible answers.

A hepatic vein **B** pulmonary artery
C mesenteric artery **D** hepatic portal vein

15 Which vessel transports blood to the small intestine?

16 Which vessel begins in a capillary bed and ends in a capillary bed?

17 The accompanying diagram shows part of the human urinary system. Which line in the following table CORRECTLY identifies the blood vessels labelled P, Q and R?

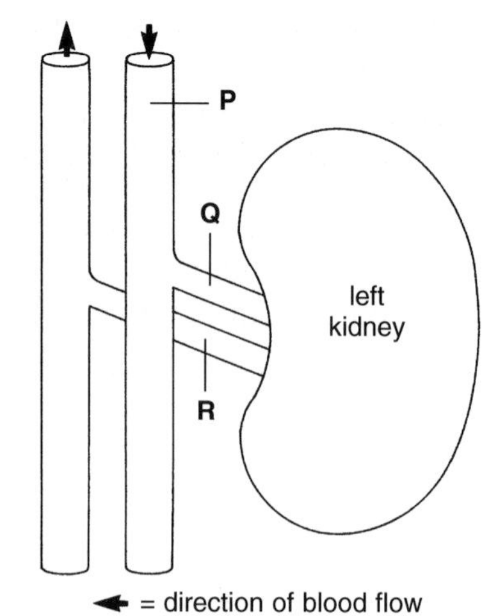

	P	Q	R
A	aorta	renal vein	renal artery
B	aorta	renal artery	renal vein
C	vena cava	renal vein	renal artery
D	vena cava	renal artery	renal vein

Questions 18 and 19 refer to the accompanying diagram of a tiny sample of human lung tissue and its blood circulatory system.

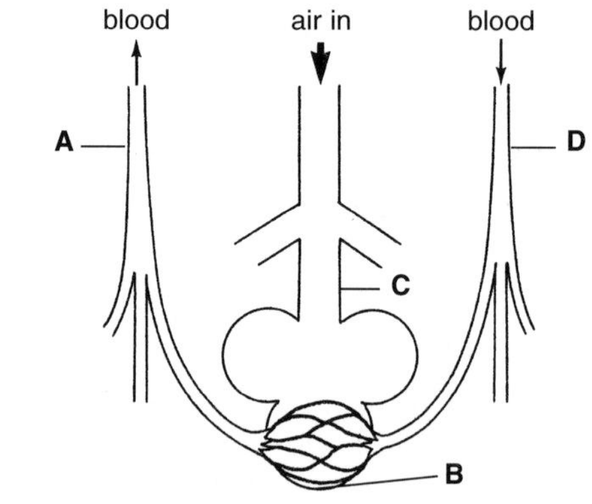

18 Which structure is a branch of the pulmonary artery?

19 Which letter indicates the region where gas exchange takes place?

20 The data in the accompanying table were obtained from an athlete before and after a 1000 metre race.

If the athlete's rate and depth of breathing remained constant during the first 5 minutes following the race, what volume of CO_2 (in cm^3) did he exhale during this period?

A 150
B 1875
C 11 250
D 225 000

	Before race	After race
Rate of breathing (breaths/min)	15	30
Average volume of each breath (cm^3)	500	1500
Concentration of CO_2 in exhaled air (%)	5	5

Items 21, 22 and 23 refer to the following information. The volumes of air entering and leaving the lungs can be measured using a piece of apparatus in which a pen moves up and down on a revolving drum making a trace as the person breathes in and out. The accompanying graph represents such a trace.

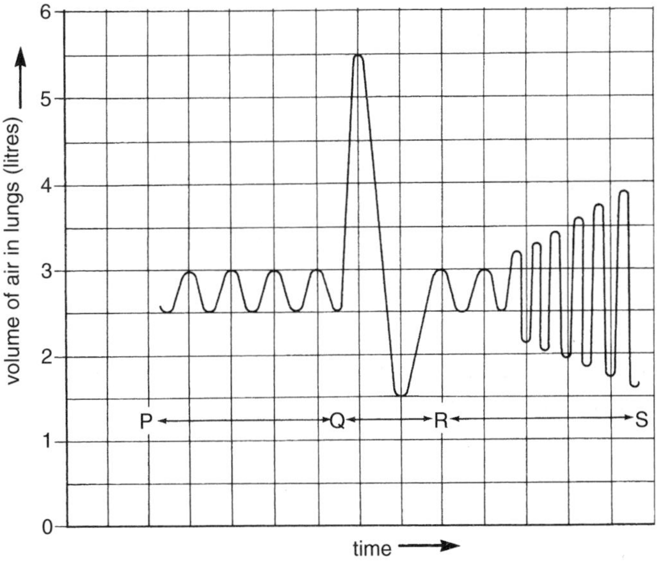

21 What volume of air (in litres) was inhaled at each breath during period P–Q when the subject was at rest?

A 0.5 **B** 1.0
C 2.5 **D** 3.0

22 The volume of air (in litres) exhaled during period Q–R was

A 4.0 **B** 7.0
C 8.0 **D** 8.5

23 Which of the following activities could account for the trace recorded during period R–S on the graph?

A nodding off to sleep
B coming to rest after exercise
C beginning an exercise programme
D inhaling the maximum volume of air and then exhaling it

Activity	Average volume of each breath (l)	Rate of breathing (breaths/min)
resting	0.5	12
jogging	1.5	24
running a race	2.5	36

24 The following data were obtained from an athlete during training.

The ratio of total volume of air breathed by the athlete per minute running a race to that breathed per minute at rest was

A 3 : 1. **B** 5 : 1. **C** 6 : 1. **D** 15 : 1.

25 Which of the following is NOT true of the body's capillary network?

A It consists of many arteries and veins in all body parts.
B It forms a dense network that serves all living cells.
C The walls of its vessels are only one cell thick.
D The combined surface area of its vessels is enormous.

Test 2

Questions 1 and 2 refer to the following possible answers.

A left ventricle **B** right ventricle
C left atrium **D** right atrium

1 Which term refers to the chamber of the heart that pumps oxygenated blood to all parts of the body?

2 Which term refers to the chamber of the heart that receives deoxygenated blood from the body?

3 In which of the accompanying diagrams of a mammalian heart, is the blood flowing in the CORRECT direction?

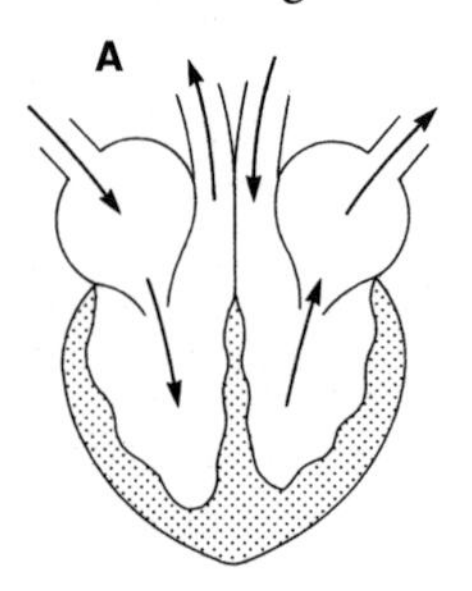

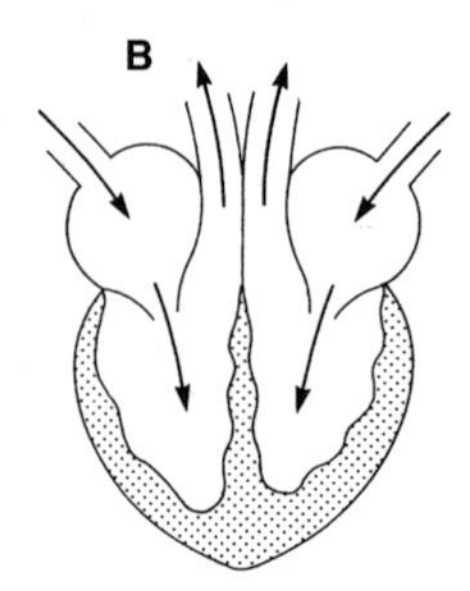

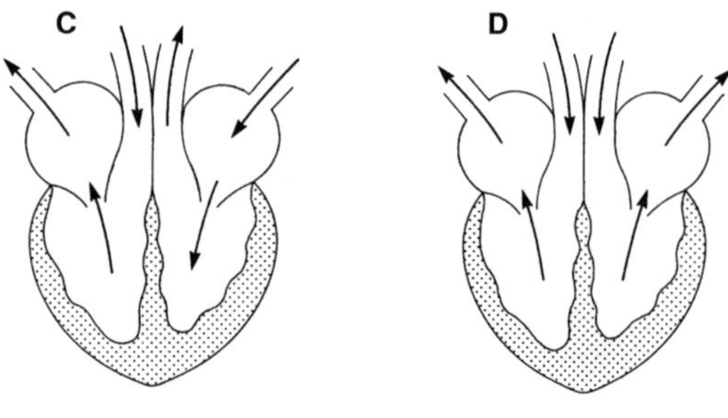

4 The accompanying diagram shows the pressure changes that occur in the left ventricle during one heart beat.

Contraction of the left ventricle is represented by the part of the graph labelled

A V–W. **B** W–X.
C X–Y. **D** Y–Z.

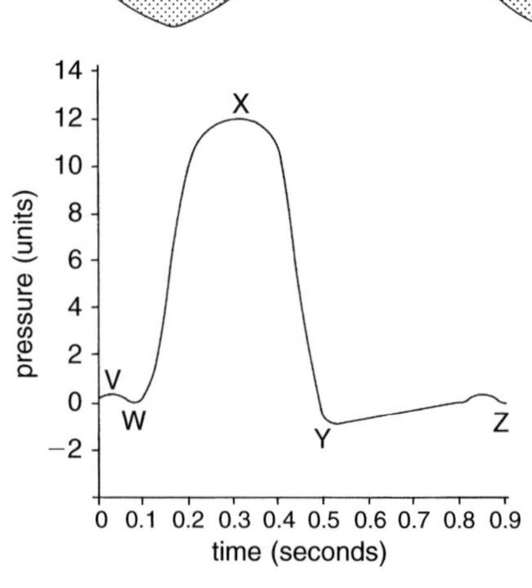

5 Which tube in the accompanying diagram of a heart CORRECTLY represents the result of a successful coronary bypass operation?

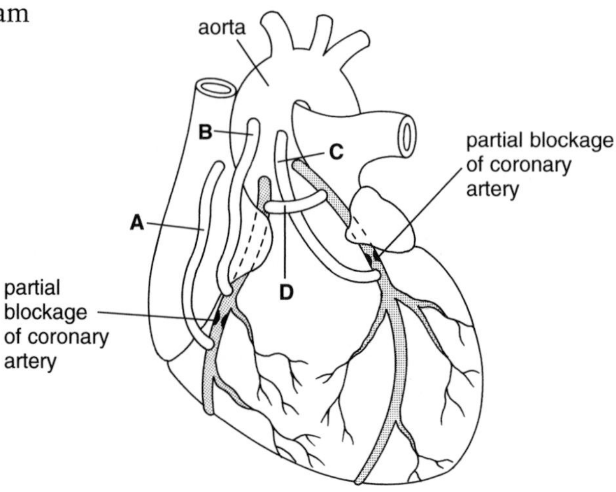

6 The accompanying graph shows the results of monitoring the heart rate of an athlete before, during and after a race over a period of 11 minutes.

During which time interval (in minutes) did the greatest increase in heart rate occur?

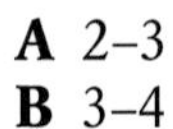

A 2–3
B 3–4
C 4–5
D 5–6

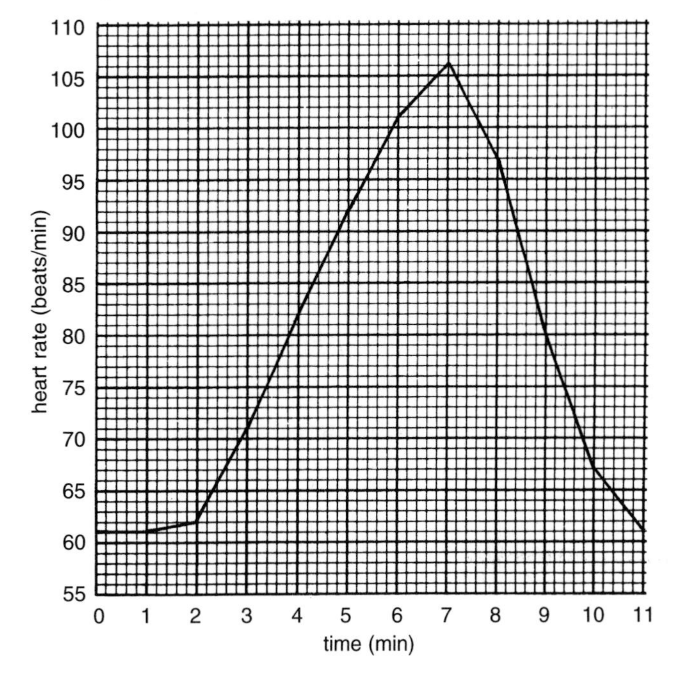

7 Which of the following differences between an artery and a vein is CORRECT?

	Feature	Artery	Vein
A	diameter of central cavity	narrow	wide
B	state of muscular wall	thin	thick
C	valves	present	absent
D	pressure of blood in vessel	low	high

8 Which of the accompanying diagrams shows the CORRECT sequence of the vessels in which blood is transported round the body?

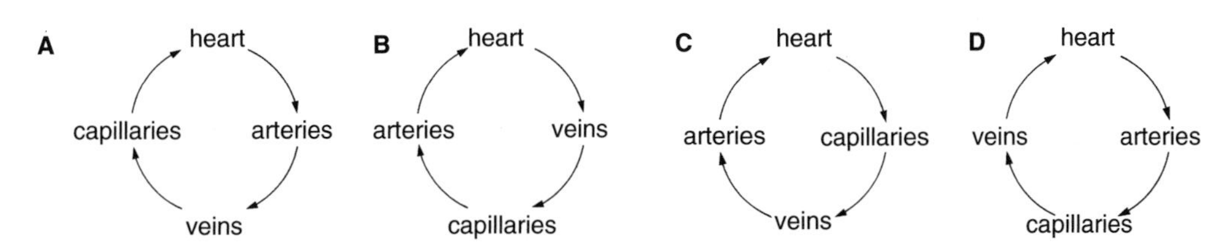

Items 9, 10, 11 and 12 refer to the accompanying diagram of the human circulatory system.

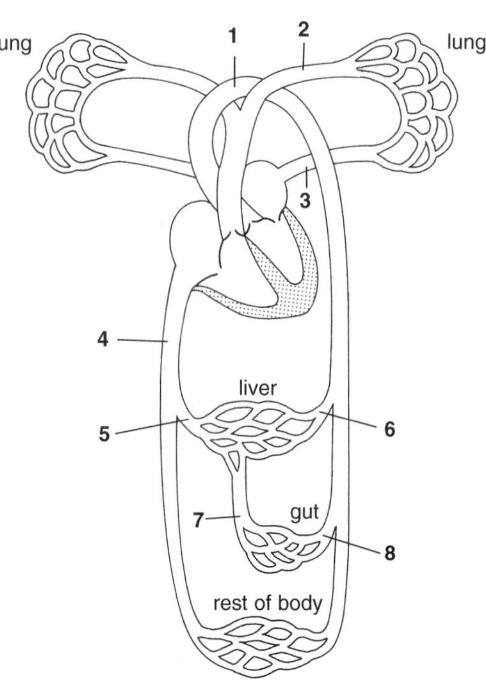

9 The vena cava is labelled

A 1. **B** 2. **C** 3. **D** 4.

10 The pulmonary artery is labelled

A 1. **B** 2. **C** 3. **D** 4.

11 The hepatic portal vein is labelled

A 5. **B** 6. **C** 7. **D** 8.

12 The mesenteric artery is labelled

A 5. **B** 6. **C** 7. **D** 8.

13 Which of the following veins contains oxygenated blood?

A hepatic **B** coronary
C pulmonary **D** hepatic portal

Items 14, 15, 16 and 17 refer to the data in the accompanying table which shows the rate of blood flow in various parts of a student's body under differing conditions of exercise.

Part of body	Rate of blood flow (cm^3/minute)		
	At rest	Light exercise	Strenuous exercise
heart muscle	250	375	750
brain	750	750	750
kidney	1100	880	605
skin	500	1500	1900
gut	1400	1100	600
skeletal muscle	1250	5000	12 500

14 The rate of blood flow remains unaltered by exercise in the

A brain. **B** kidney. **C** skin. **D** gut.

15 In which part of the body does rate of blood flow decrease as exercise becomes more strenuous?

A heart muscle **B** brain **C** kidney **D** skin

16 Compared with the situation during light exercises, rate of blood flow in skeletal muscle during strenuous exercise increases by a factor of

A 0.4. **B** 2.5. **C** 4.0. **D** 7500.0.

17 Compared with the situation at rest, the rate of blood flow in the kidney during light exercise has decreased by

A 2.2% **B** 20.0% **C** 22.0% **D** 25.0%

18 Which of the following gives the CORRECT order of the structures through which air passes on being inhaled?

A trachea, bronchus, bronchiole, alveolus
B bronchus, trachea, alveolus, bronchiole
C trachea, bronchus, alveolus, bronchiole
D bronchus, trachea, bronchiole, alveolus

19 Which of the accompanying diagrams of gas exchange in a pair of air sacs in a human lung is CORRECT?

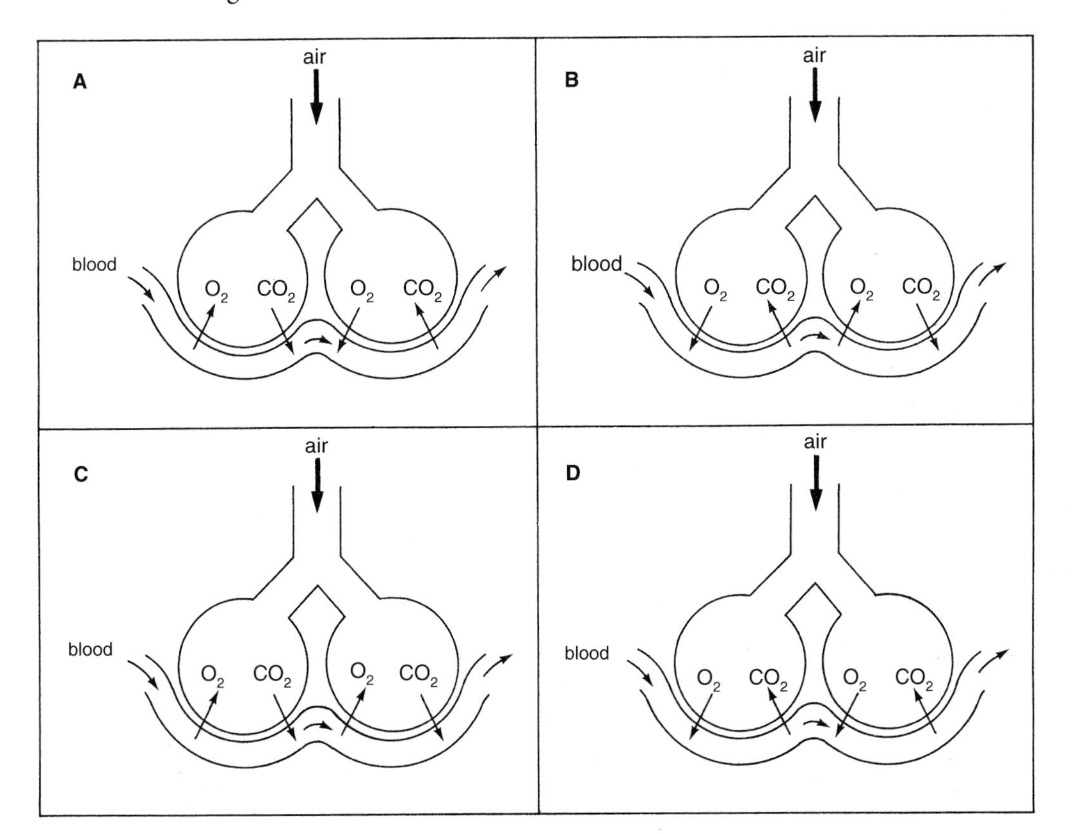

20 Various aspects of a woman's breathing were investigated. The results are shown in the accompanying table.

Concentration of oxygen in inhaled air (%)	20
Concentration of oxygen in exhaled air (%)	16
Number of breaths per minute	15
Average volume of each breath (cm^3)	500

The volume of oxygen (in cm^3) absorbed by her lungs each minute was

A 20. **B** 300. **C** 1200. **D** 1500.

Questions 21 and 22 refer to the accompanying graph.

21 At what point on the graph is a shallow breath being exhaled?

22 At what point on the graph is a deep breath being inhaled?

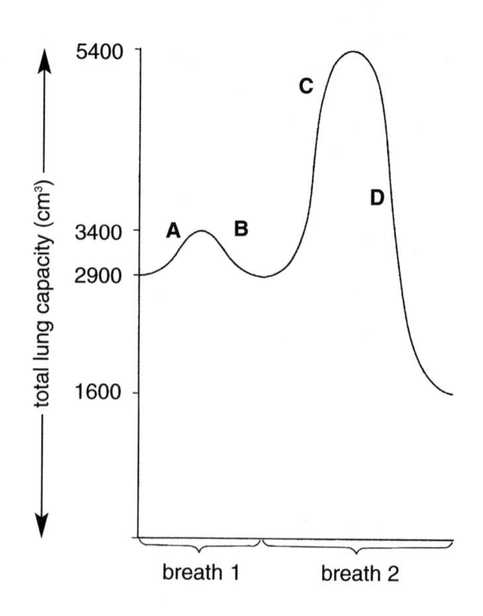

Items 23 and 24 refer to the accompanying graph which shows the relationship between breathing and the concentration of carbon dioxide in inhaled air.

23 When the carbon dioxide concentration in inhaled air is increased from 2% to 6%, the average depth of breathing (in cm^3) increases by

A 12.4. **B** 13.0.
C 1240.0. **D** 1400.0.

24 As the carbon dioxide concentration in inhaled air increases

A rate and depth of breathing increase at the same rate.
B rate of breathing shows an increase before depth of breathing.
C rate and depth of breathing begin to increase at the same concentration of CO_2.
D depth of breathing shows an increase before rate of breathing.

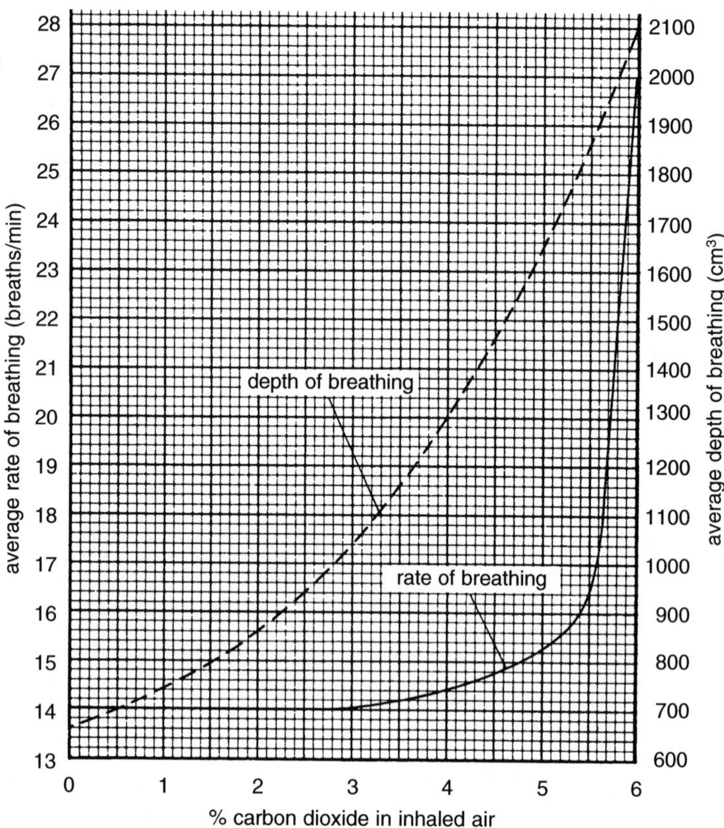

25 In which of the following groups do ALL THREE features contribute to an air sac's efficiency as a structure of gas exchange?

A moist surface, thin lining and large surface area
B moist surface, thick lining and large surface area
C dry surface, thin lining and large surface area
D moist surface, thin lining and small surface area

14 Blood

Test 1

Items 1, 2, 3, 4 and 5 refer to the accompanying diagram of a tiny part of a blood capillary in a tissue.

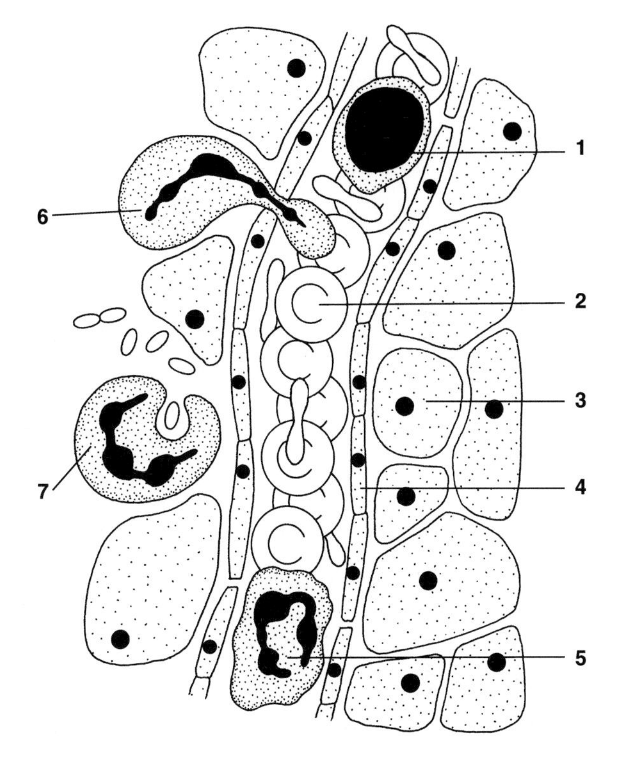

1 Which cell is a lymphocyte?

A 1 **B** 3 **C** 4 **D** 7

2 Which cell is a red blood cell?

A 1 **B** 2 **C** 3 **D** 7

3 Which two cells are phagocytes?

A 1 and 5 **B** 1 and 6
C 2 and 7 **D** 6 and 7

4 Comparison of cell types 2 and 3 reveals that only type 2 has

A a cell membrane.
B cytoplasm.
C a nucleus.
D haemoglobin.

5 Comparison of cell types 2 and 4 reveals that only type 4 has

A a cell membrane.
B cytoplasm.
C a nucleus.
D haemoglobin.

6 Which of the following are BOTH carried round the body dissolved in blood plasma?

A glucose and amino acids
B white blood cells and glucose
C amino acids and red blood cells
D red blood cells and white blood cells

7 Oxygen needed for tissue respiration is carried round the body in the bloodstream. Almost all of this oxygen is transported by the

A blood plasma.
B red blood cells.
C white blood cells and blood plasma.
D red blood cells and white blood cells.

8 The presence of large quantities of acid in the bloodstream would lead to problems since blood functions best at pH

A 6.4. **B** 7.4. **C** 8.4. **D** 9.4.

9 Carbon dioxide produced during tissue respiration is transported in the bloodstream by

A blood plasma only.
B red blood cells only.
C red blood cells and blood plasma.
D red blood cells and white blood cells.

Items 10 and 11 refer to the following possible answers.

A association **B** phagocytosis
C dissociation **D** antibody formation

10 What name is given to the process by which oxyhaemoglobin off-loads its oxygen when the oxygen concentration in the surroundings is low?

11 What name is given to the immune response made by lymphocytes in the presence of a disease-causing organism?

12 Which equation is CORRECT?

A haemoglobin + oxygen $\xrightarrow[\text{in lungs}]{\text{association}}$ oxyhaemoglobin

B haemoglobin + oxygen $\xrightarrow[\text{in tissues}]{\text{dissociation}}$ oxyhaemoglobin

C oxyhaemoglobin $\xrightarrow[\text{in lungs}]{\text{dissociation}}$ haemoglobin + oxygen

D oxyhaemoglobin $\xrightarrow[\text{in tissues}]{\text{association}}$ haemoglobin + oxygen

Items 13 and 14 refer to the accompanying graph.

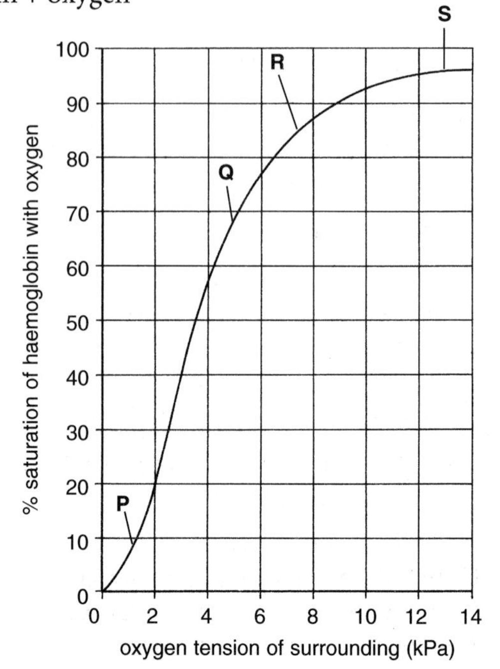

13 If haemoglobin, already acclimatised to surroundings of oxygen tension 12 kPa, was transported to body tissues of oxygen tension 3 kPa, it would off-load oxygen. The resulting decrease in its percentage saturation with oxygen would be

A 9. **B** 40. **C** 55. **D** 95.

14 The oxygen tension of the blood present in the aorta is indicated on the graph by letter

A P. **B** Q. **C** R. **D** S.

15 Which of the following differences between red and white blood cells from a human being is CORRECT?

	Feature	Red cell	White cell
A	shape	irregular	biconcave
B	number	large	small
C	nucleus	present	absent
D	function	protection	oxygen transport

16 During phagocytosis, a trapped bacterium is destroyed by

A enzymes contained in lysosomes.
B antibodies contained in lymphocytes.
C enzymes contained in lymphocytes.
D antibodies contained in lysosomes.

17 Which of the accompanying diagrams shows the structure of an antibody molecule? (R = receptor site, N = non-receptor site.)

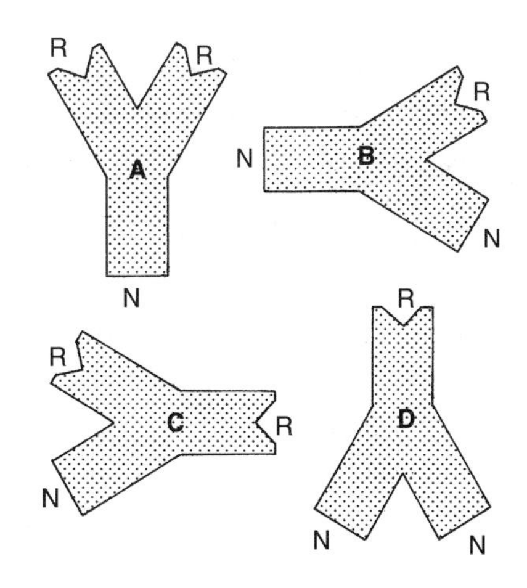

18 The following steps occur during one of the human body's defence mechanisms.

1 antibodies are released into the bloodstream
2 virus gains access to the body and multiplies
3 antigens combine with antibodies forming harmless complexes
4 lymphocytes respond by producing antibodies

The correct sequence of these stages is

A 2, 1, 4, 3. **B** 2, 4, 1, 3.
C 4, 2, 1, 3. **D** 4, 2, 3, 1.

19 The accompanying diagram shows a virus (V), the cause of a fatal disease, and a weakened version of the virus (W).

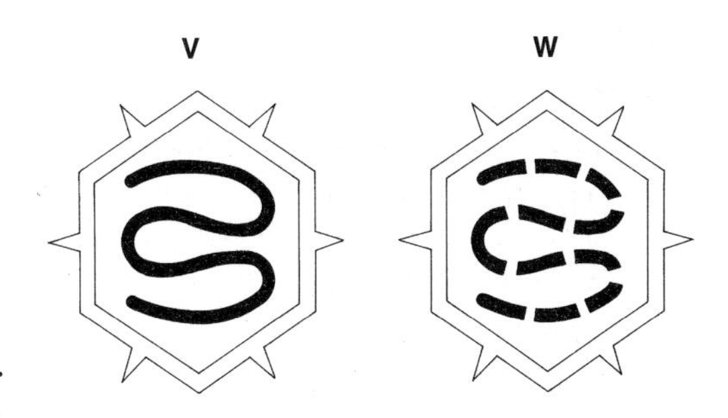

During an immunisation programme, patients would receive

A V and acquire immunity by natural means.
B V and acquire immunity by artificial means.
C W and acquire immunity by natural means.
D W and acquire immunity by artificial means.

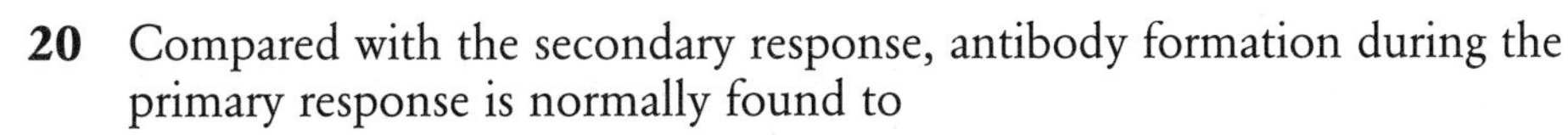

20 Compared with the secondary response, antibody formation during the primary response is normally found to

A reach a higher level.
B be maintained for a longer time.
C be unable to prevent the disease.
D occur at a much more rapid rate.

Test 2

Items 1, 2 and 3 refer to the following diagram which classifies some of the components human blood.

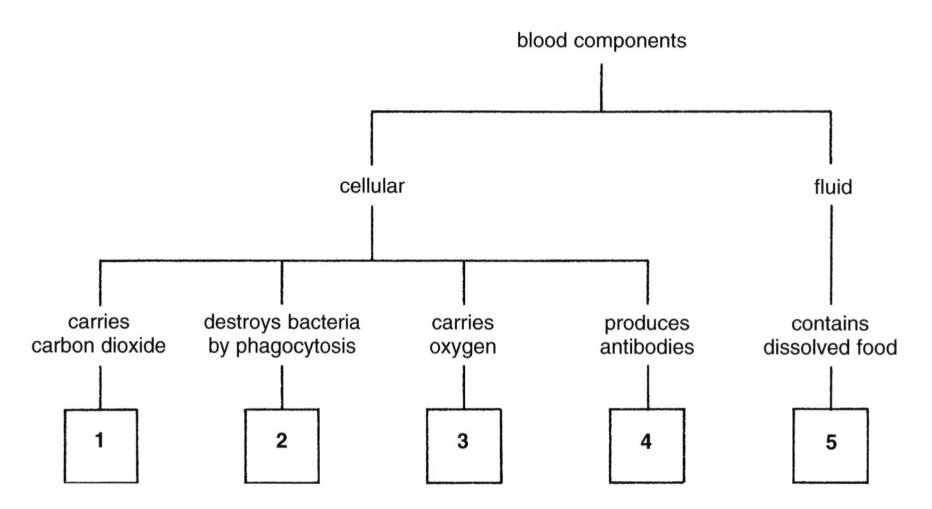

1 'Plasma' should have been written in box

A 1. **B** 3. **C** 4. **D** 5.

2 'Red blood cells' should have been written in boxes

A 1 and 3. **B** 1 and 5.
C 2 and 3. **D** 2 and 4.

3 'Lymphocytes' should have been written in box

A 1. **B** 2. **C** 3. **D** 4.

4 The quantity of carbon dioxide that can be carried dissolved in blood plasma is limited by the fact that carbon dioxide and water combine to form an

A acid. **B** alkali.
B antigen. **D** antibody.

5 Which feature of a red blood cell gives it a large surface area in relation to its volume?

A its flexibility
B its biconcave shape
C the lack of a nucleus
D the presence of haemoglobin

Questions 6 and 7 refer to the following possible answers.

A association **B** phagocytosis
C dissociation **D** antibody production

6 What name is given to the process by which haemoglobin loads up with oxygen when the oxygen concentration in the surroundings is high?

7 What name is given to the process by which macrophages engulf and destroy bacteria?

8 Which equation is CORRECT?

A haemoglobin + oxygen $\xrightarrow[\text{in lungs}]{\text{dissociation}}$ oxyhaemoglobin

B haemoglobin + oxygen $\xrightarrow[\text{in tissues}]{\text{association}}$ oxyhaemoglobin

C oxyhaemoglobin $\xrightarrow[\text{in tissues}]{\text{dissociation}}$ haemoglobin + oxygen

D oxyhaemoglobin $\xrightarrow[\text{in lungs}]{\text{association}}$ haemoglobin + oxygen

Items 9, 10, 11 and 12 refer to the following information. At high altitudes, the air is thinner and less oxygen is gained by the body per breath. Graphs 1 and 2 refer to the altitudes reached by and the red blood cell counts of a group of climbers on a mountaineering expedition.

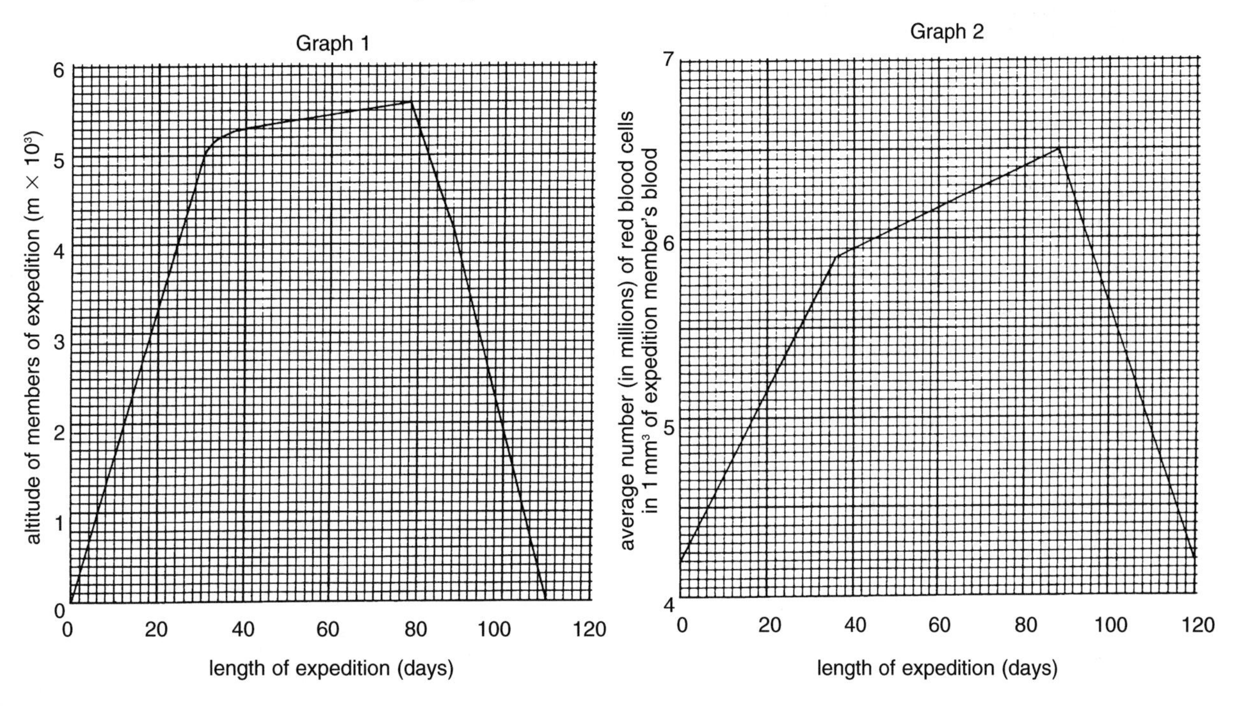

9 Which line in the following table gives the CORRECT values for day 24?

	Altitude reached ($m \times 10^3$)	Number of red blood cells (millions per mm^3)
A	4.0	5.35
B	4.7	5.55
C	4.0	5.55
D	4.7	5.35

10 On which day did the climbers reach the highest altitude?

A 78 **B** 79 **C** 84 **D** 88

11 How many days did the members of the expedition spend at an altitude of 5×10^3 m or above?

A 46 **B** 51 **C** 52 **D** 56

12 What was the average number of red blood cells, in millions per mm^3, present in a climber's blood on day 84?

A 4.20 **B** 4.70 **C** 6.45 **D** 6.50

13 One cubic millimetre of a patient's blood was found to contain 8898 white blood cells and 5 338 800 red blood cells. Which of the following expresses these figures correctly as a ratio?

A 60 white : 1 red
B 60 red : 1 white
C 600 white : 1 red
D 600 red : 1 white

14 Ideal conditions for growth of disease-causing bacteria in the human body are provided by a supply of

A food, carbon dioxide and warmth.
B warmth, moisture and food.
C moisture, food and carbon dioxide.
D carbon dioxide, moisture and warmth.

15 The accompanying diagram shows four of the stages that occur during the process of phagocytosis.

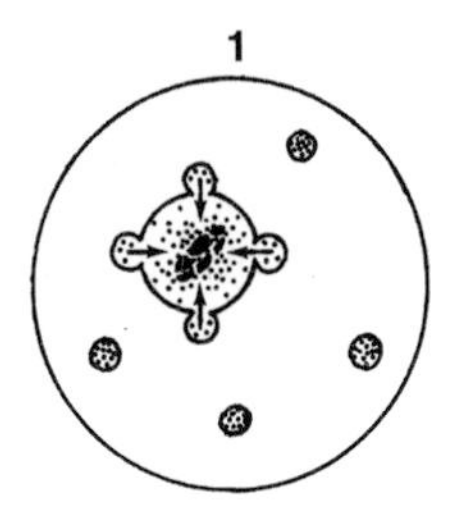

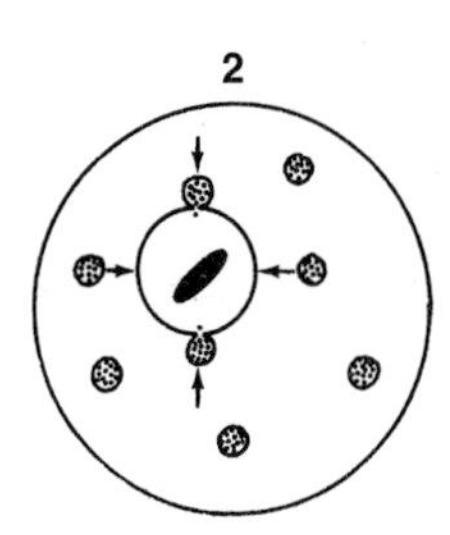

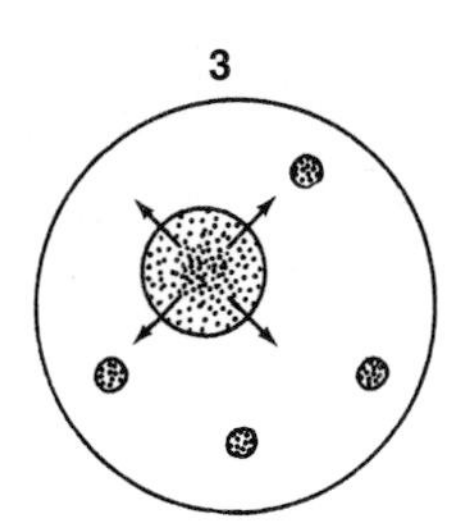

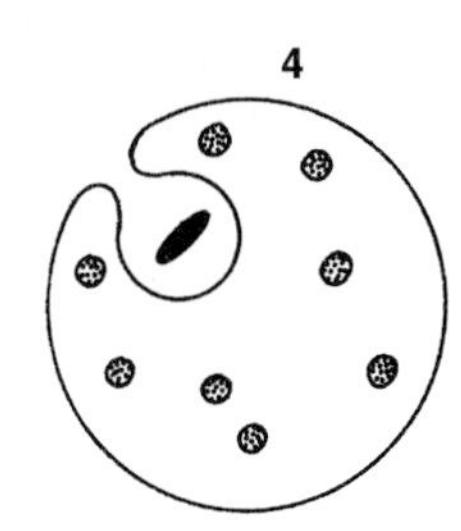

Which set in the table CORRECTLY matches each of these numbered stages with its description?

Description of state	**Set**			
	A	**B**	**C**	**D**
Products of digestion pass into cytoplasm of phagocyte	1	3	4	2
Some lysosomes move towards and fuse with vacuole	4	2	1	3
Phagocyte forms vacuole around bacterium	2	4	3	1
Digestive enzymes break down bacterium	3	1	2	4

16 Pus consists largely of dead

A bacteria and lymphocytes.
B antibodies and phagocytes.
C bacteria and phagocytes.
D antibodies and lymphocytes.

17 The first of the two accompanying diagrams shows a group of antibodies.

Which of the four viruses shown in the second diagram could be rendered harmless by these antibodies binding with its antigens?

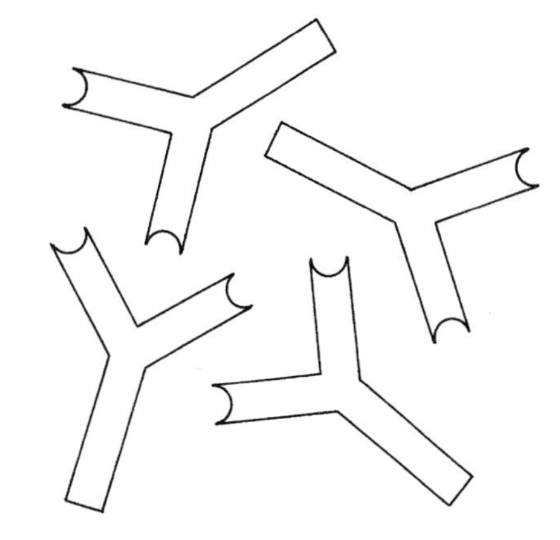

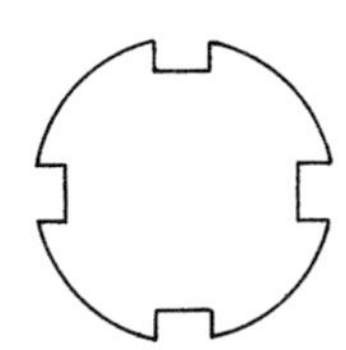

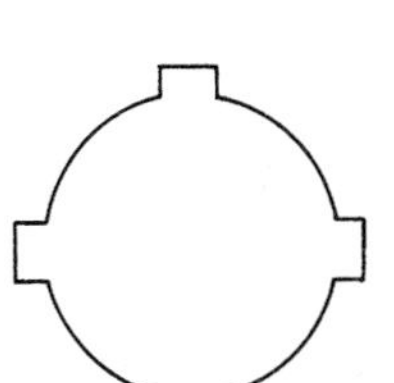

18 Which line in the following table is CORRECT?

	Phagocytosis	**Antibody formation**
A	specific	specific
B	non-specific	specific
C	specific	non-specific
D	non-specific	non-specific

19 Which of the following graphs best represents the body's primary and secondary responses on exposure to a disease-causing micro-organism?

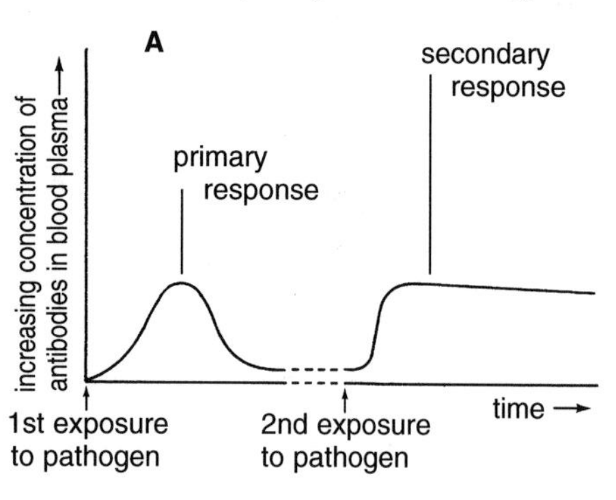

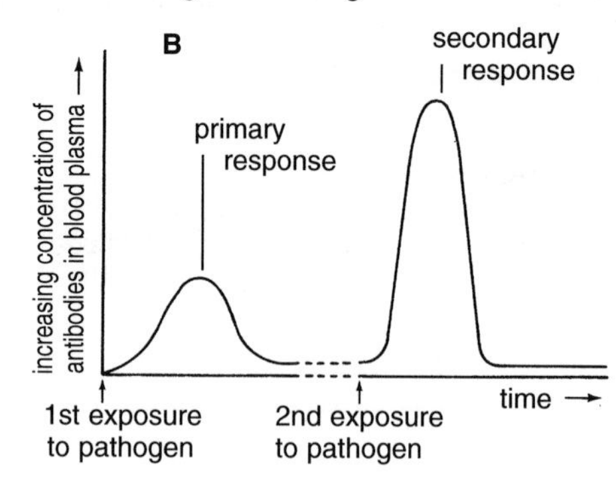

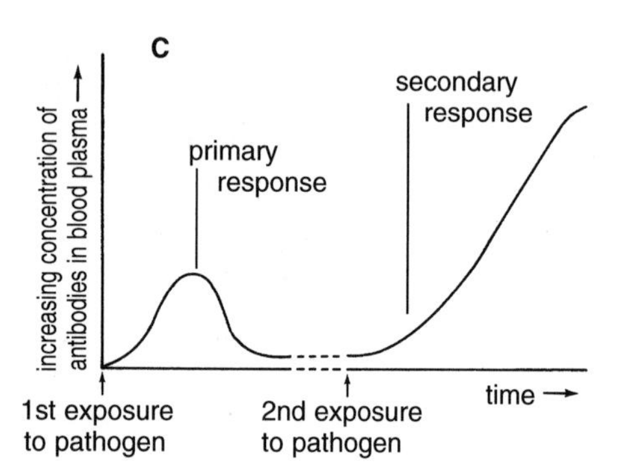

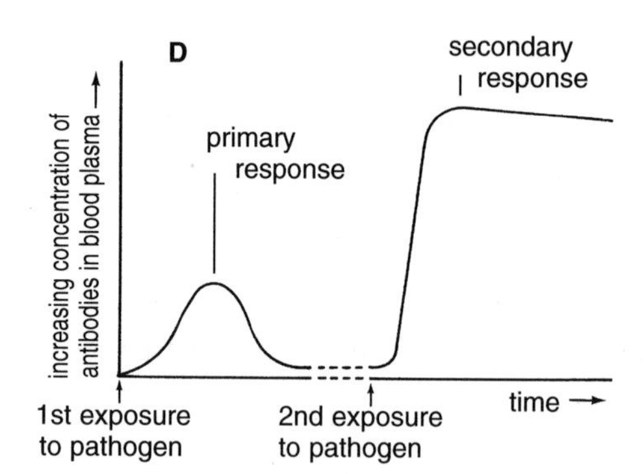

20 The following list gives three case histories involving a disease caused by a certain micro-organism.

Patient 1 suffered the disease without medical treatment.
Patient 2 received vaccine long before coming into contact with the micro-organism.
Patient 3 consumed a course of antibiotics on becoming ill with the disease.

All three patients survived. Who now possesses memory cells capable of producing antibody-forming cells in the event of a second exposure to the micro-organism?

A 1 only. **B** 1 and 2 only.
C 2 and 3 only. **D** 1, 2 and 3.

15 Brain and nervous system

Test 1

Items 1, 2, 3, 4, 5, 6 and 7 refer to the accompanying diagram of the human brain.

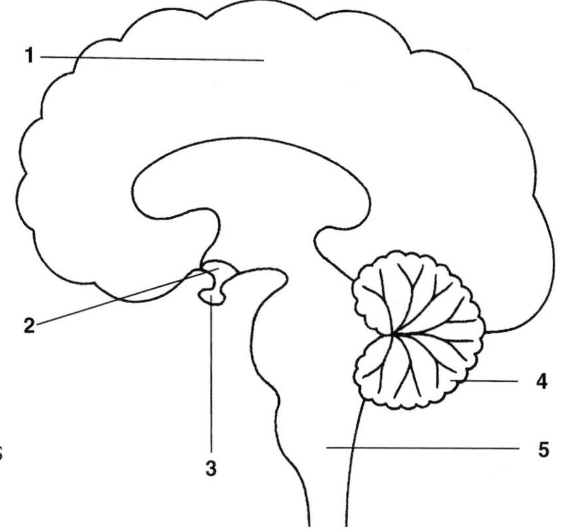

1 The cerebral hemisphere is

A 1. **B** 2. **C** 4. **D** 5.

2 The medulla is

A 2. **B** 3. **C** 4. **D** 5.

3 The cerebellum is

A 1. **B** 3. **C** 4. **D** 5.

4 The hypothalamus is

A 2. **B** 3. **C** 4. **D** 5.

5 The region of the brain that contains the centre responsible for regulating heart rate is

A 1. **B** 3. **C** 4. **D** 5.

6 The region of the brain that controls muscular co-ordination is

A 2. **B** 3. **C** 4. **D** 5.

7 The region of the brain responsible for higher mental faculties such as reasoning and imagination is

A 1. **B** 2. **C** 4. **D** 5.

Questions 8 and 9 refer to the accompanying diagram of the human cerebrum.

8 Which is the motor area?

9 Which is the sensory area?

10 On being asked and agreeing to sign your name, information passes through certain parts of your body in the order

A receptors in ear → motor area in brain → sensory area in brain → muscles in fingers
B receptors in ear → sensory area in brain → motor area in brain → muscles in fingers
C sensory area in brain → receptors in ear → muscles in fingers → motor area in brain
D receptors in ear → sensory area in brain → muscles in fingers → motor area in brain

A
B
C
D

11 Effectors are usually

A nerves. **B** muscles.
C grey matter. **D** sense organs.

12 Which of the following diagrams shows the CORRECT sequence of neurones in a reflex arc?

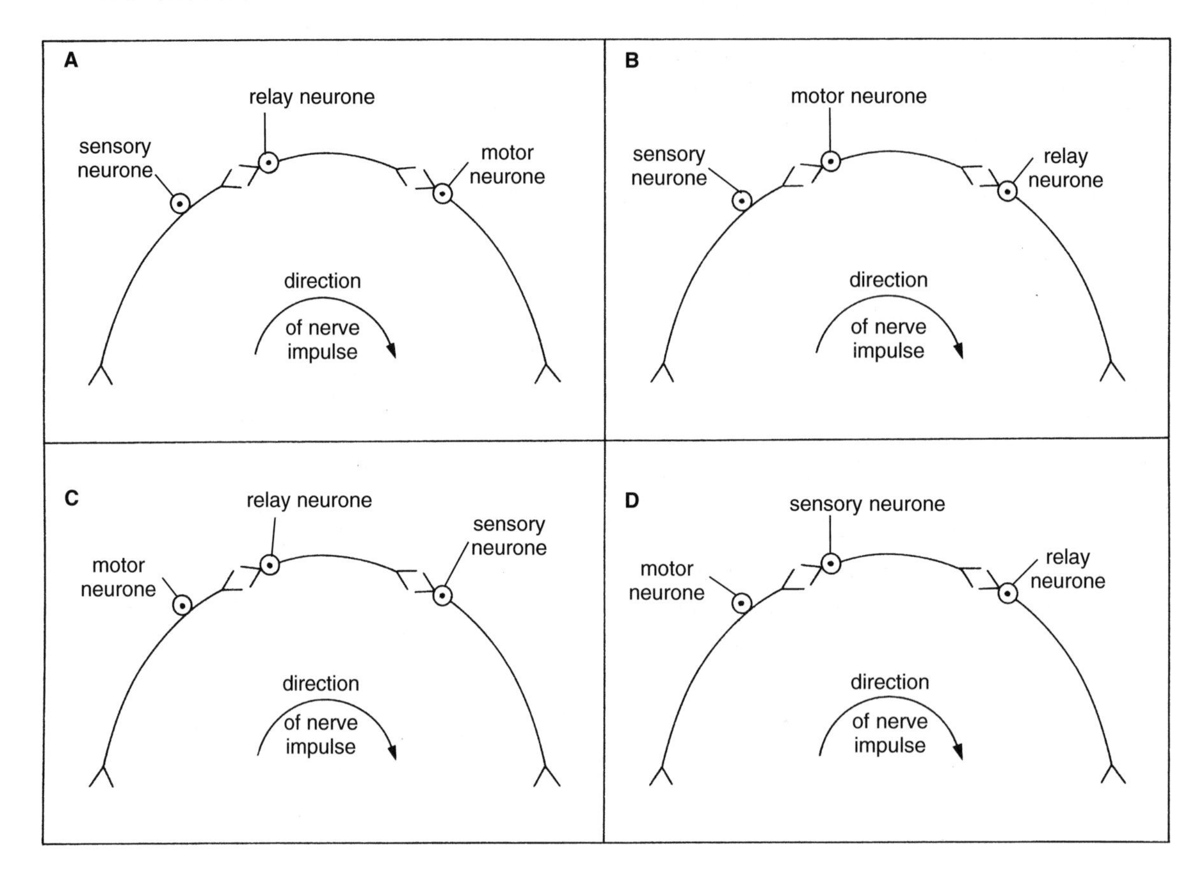

13 A reflex action is a rapid,

A voluntary response to a stimulus.
B involuntary response to a stimulus.
C voluntary stimulus to a response.
D involuntary stimulus to a response.

14 Which of the following would NOT result in a reflex action?

A grit being blown onto the surface of your eye
B chewed food being passed to the back of your mouth✓
C pollen grains being inhaled into your nasal tract
D chocolates being offered to you by a visiting friend

Questions 15 and 16 refer to tables 1 and 2 that follow.

Table 1

Reflex action	Stimulus	Response	Protective function
dilation of eye pupil	P	contraction of certain iris muscles	Q
constriction of eye pupil	R	contraction of certain iris muscles	S

Table 2

	Stimulus	Protective function
A	dim light	prevents damage to eye
B	dim light	improves vision in poor lighting
C	bright light	prevents damage to eye
D	bright light	improves vision in poor lighting

15 Which line in Table 2 provides the answers to blank boxes P and Q in Table 1?

16 Which line in Table 2 provides the answers to blank boxes R and S in Table 1?

17 The accompanying graph shows the effect of increasing external temperature on the temperature of four parts of the human body.

Which line represents the temperature of the rectum?

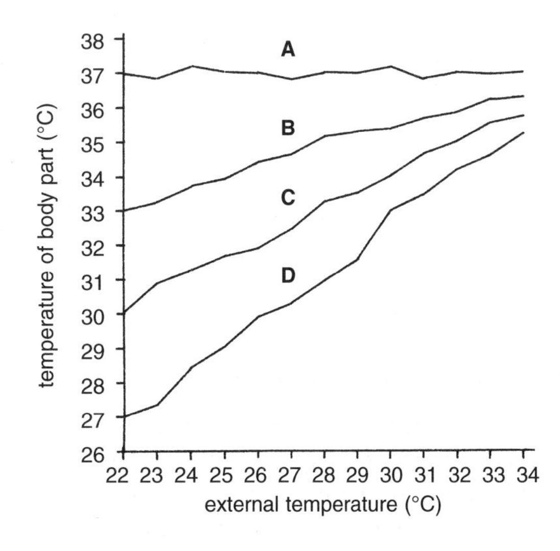

18 Which of the following pairs of corrective mechanisms would occur in response to an increase in body temperature?

A constriction of blood capillaries in the skin and contraction of hair erector muscles
B constriction of blood capillaries in the skin and relaxation of hair erector muscles
C dilation of blood capillaries in the skin and contraction of hair erector muscles
D dilation of blood capillaries in the skin and relaxation of hair erector muscles

Items 19, 20 and 21 refer to the accompanying diagram of temperature regulation in the human body.

normal body temperature

increase in body temperature

decrease in body temperature

detection centre

corrective mechanism type 1

corrective mechanism type 2

normal body temperature

19 The location of the detection centre is in the

A skin. **B** body shell.
C hypothalamus. **D** pituitary gland.

20 An example of corrective mechanism type 1 is

A shivering.
B increased sweating.
C increased metabolic rate.
D contraction of hair erector muscles.

21 Which of the following correctly describes an example of corrective mechanism type 2?

A vasoconstriction which leads to an increased volume of blood flowing through the skin
B vasodilation which leads to an increased volume of blood flowing through the skin
C vasoconstriction which leads to a decreased volume of blood flowing through the skin
D vasodilation which leads to a decreased volume of blood flowing through the skin

22 Which of the following is a voluntary mechanism of temperature regulation in response to overcooling?

A decreased production of sweat
B increased rate of metabolism
C shivering uncontrollably
D exercising vigorously

Items 23, 24 and 25 refer to the accompanying diagram which records a hospital patient's body temperature taken over a period of five days using a clinical thermometer.

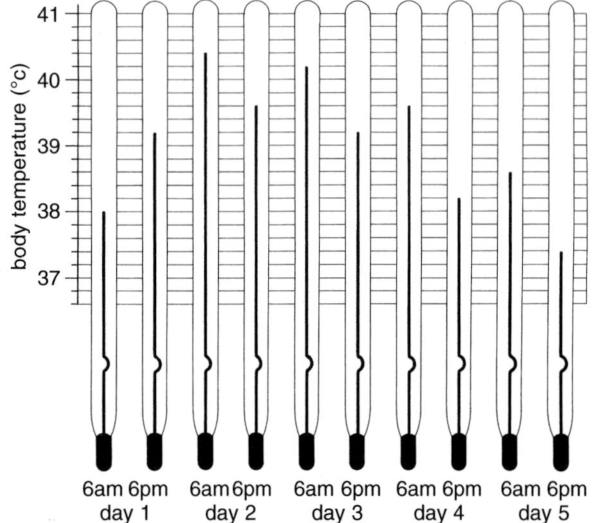

23 The greatest change in temperature occurred between

A 6am and 6pm on day 1.
B 6pm on day 1 and 6am on day 2.
C 6am and 6pm on day 4.
D 6am and 6pm on day 5.

24 During which of the following periods of time was the patient sweating profusely but her body temperature was NOT corrected by this negative feedback mechanism?

Between 6am and 6pm on day

A 1. **B** 2. **C** 3. **D** 4.

25 During which of the following periods of time was the patient's skin in a state of vasodilation and her body temperature was corrected by this negative feedback mechanism?

A between 6pm on day 2 and 6am on day 3
B between 6pm on day 3 and 6am on day 4
C between 6pm on day 4 and 6am on day 5
D between 6am and 6pm on day 5

Test 2

1 In the accompanying diagram of part of the human brain, the structures belonging to the central nervous system are numbered

A 1 and 2. **B** 2 and 3.
C 1 and 4. **D** 3 and 4.

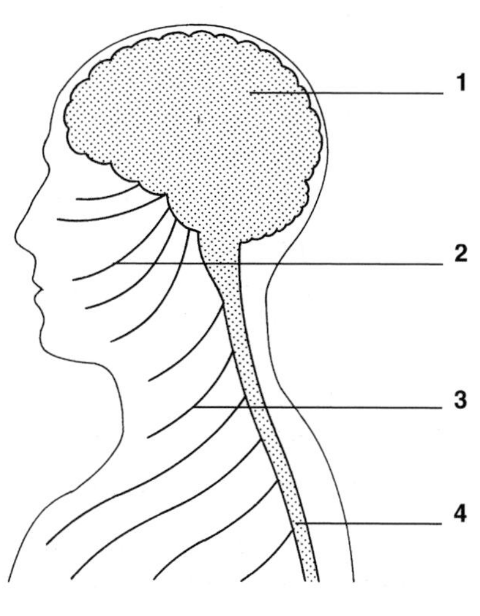

Items 2, 3, 4 and 5 refer to the following possible answers.

A cerebrum **B** cerebellum
C medulla **D** hypothalamus

2 Which part of the brain controls balance?

3 Which part of the brain is being employed by you to think about the answer to this question?

4 Which region of the brain contains a regulatory centre that controls the body's water balance?

5 Which region of the brain contains the centre that regulates rate of breathing?

6 The accompanying diagram shows the left cerebral hemisphere of the human brain.

Which region, on receiving messages from other parts of the brain, sends motor impulses to skeletal muscles bringing about movement?

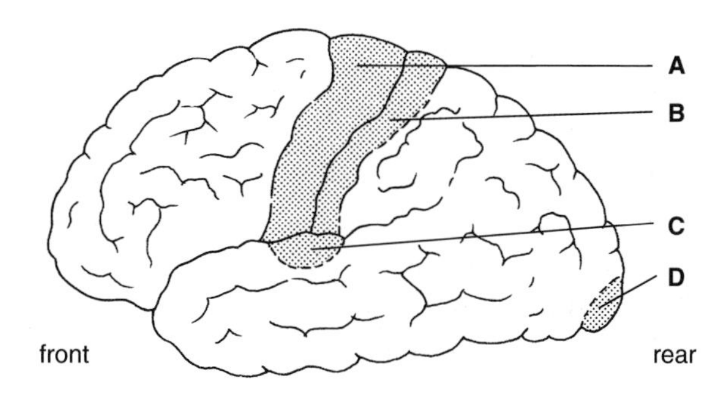

7 When a certain region of the brain is affected by excessive consumption of alcohol, the person's movements become clumsy and uncoordinated. The affected part of the brain is the

A medulla.
B pituitary.
C cerebellum
D hypothalamus.

Questions 8, 9 and 10 refer to the two accompanying diagrams. The first indicates the share of the brain's sensory area allocated to each body part; the second shows an imaginary human figure ('sensory homunculus') whose body parts have been drawn in proportion to their sensitivity as opposed to their actual size.

8 Which of the following is a large part of a normal human body yet is represented by a relatively small part of the cerebrum's sensory region?

A leg
B face
C hand
D tongue

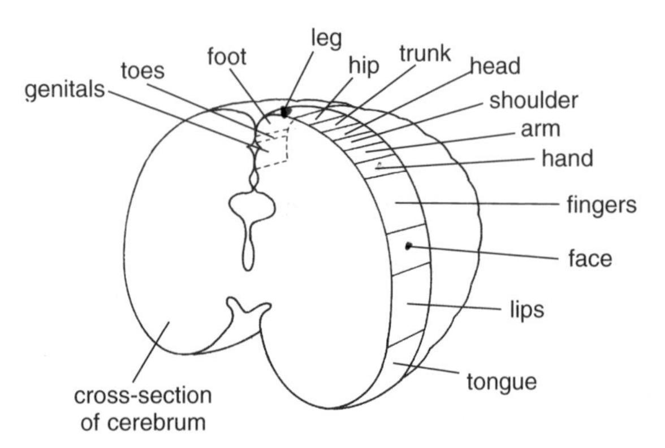

9 Which of the following body parts contains most sensory receptors relative to its actual size?

A arm
B face
C trunk
D shoulder

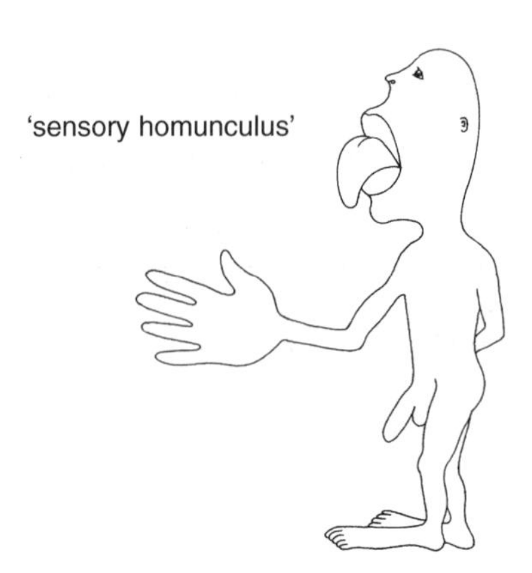

10 Which of the following structures would have fewest sensory nerve endings in relation to its actual size?

A hip
B lip
C tongue
D finger

11 Which of the following diagrams CORRECTLY represents the flow of information through the nervous system?

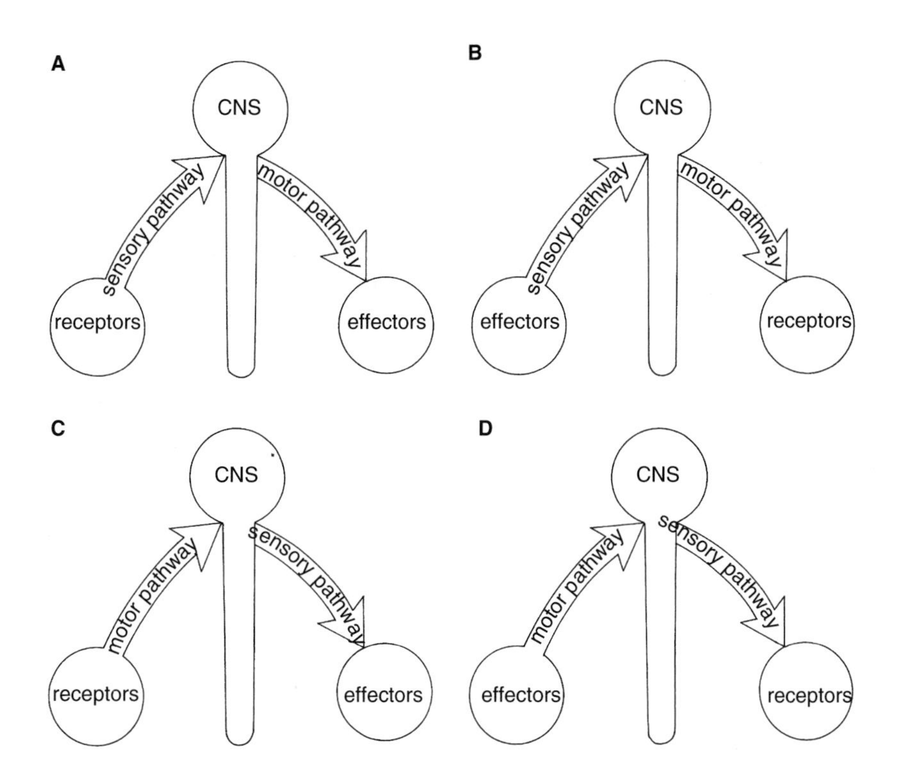

Items 12, 13 and 14 refer to the following diagram of a simple reflex arc.

12 Which lettered structure indicates a sensory nerve fibre?

13 Which lettered structure represents a motor nerve fibre?

14 The impulse passing along route X could be going to the

A heart. **B** cerebrum.
C pituitary. **D** hypothalamus.

15 Which of the following statements refers to a reflex action?

A The stimulus is originally picked up by a relay neurone.
B The nerve impulse is passed from the motor to the sensory neurone.
C The nerve impulse always passes from receptor to effector via the brain.
D The effector responds on receiving nerve impulses from a motor neurone.

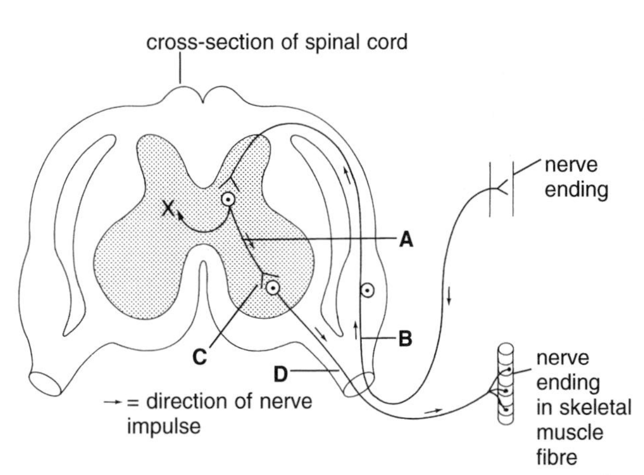

Questions 16, 17, 18 and 19 refer to the accompanying table.

Reflex action	Stimulus	Response	Protective function
A	harmful object approaching eye surface	contraction of eyelid muscle	prevents damage to eye
B	presence of food in gut	contraction of muscle in gut wall	ensures movement and efficient digestion of food
C	heat from a naked flame	contraction of flexor muscle	moves limb to safety
D	foreign particles in nasal tract	sudden contraction of chest muscles	removes unwanted particles from nose

16 Which line in the table refers to sneezing?

17 Which line in the table refers to blinking?

18 Which line in the table refers to peristalsis?

19 Which line in the table refers to limb withdrawal?

20 A thermoreceptor is a structure which

A effects a series of involuntary contractions of skeletal muscle.
B secretes liquid onto the skin surface giving a cooling effect.
C brings about a negative feedback control mechanism.
D detects changes in the temperature of the body.

21 Which line in the following table is CORRECT?

	Region of body	Examples of components	Temperature of region (°C)
A	core	skin and skeletal muscles	33
B	core	heart and brain	37
C	shell	skin and skeletal muscles	37
D	shell	heart and brain	33

22 Which diagram represents vasodilation of blood vessels in the skin and the role that this process plays in temperature regulation?

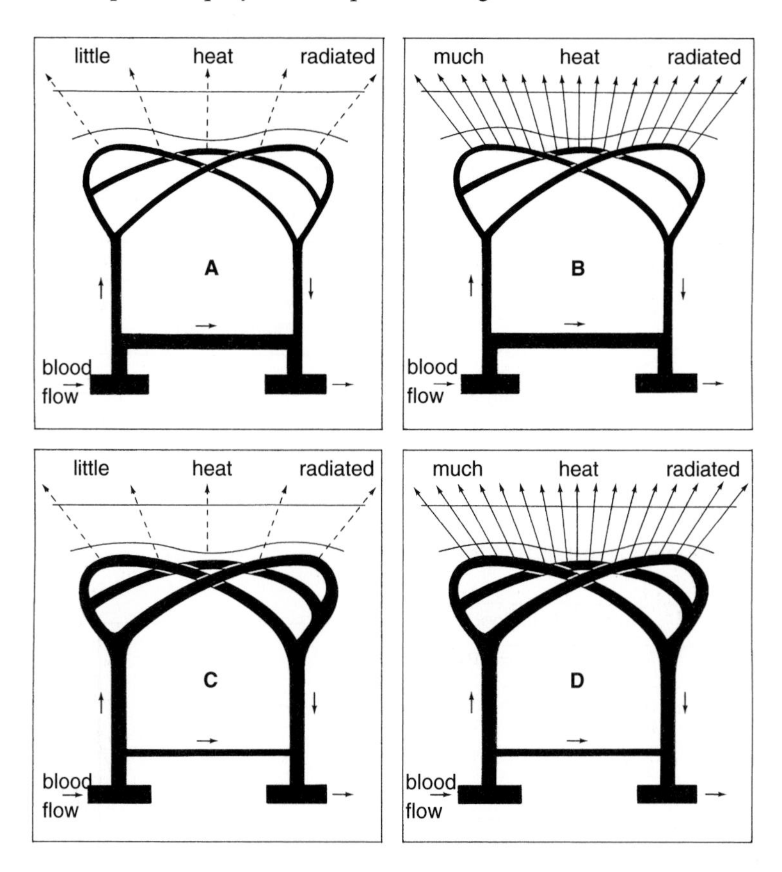

23 Which of the following diagrams illustrates an attempt by the human body to correct overcooling?

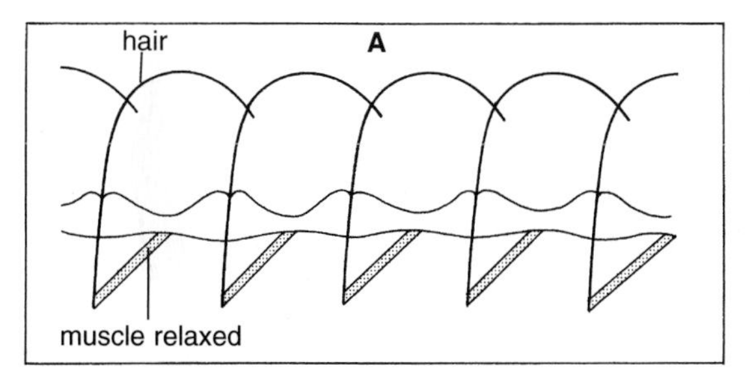

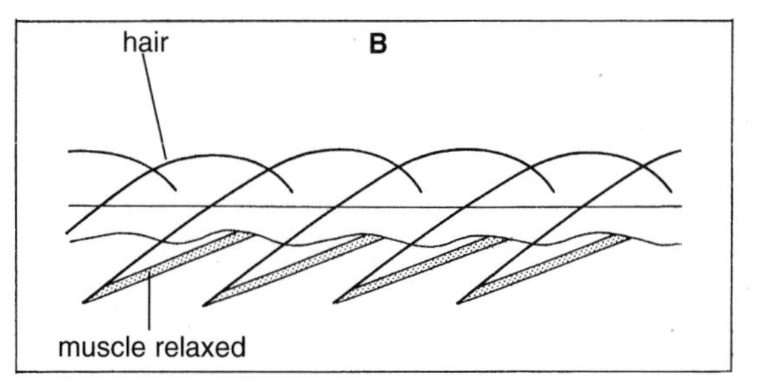

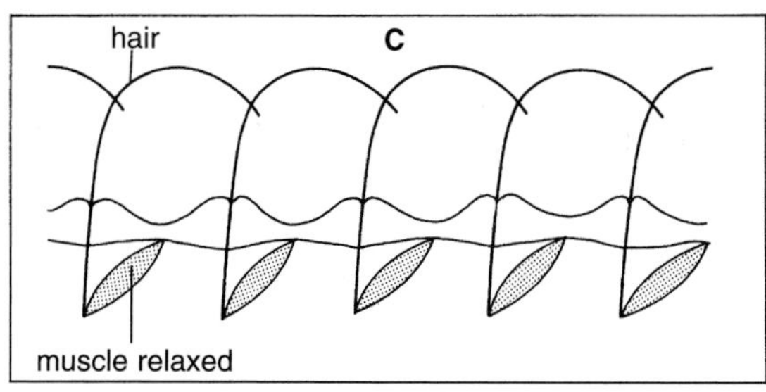

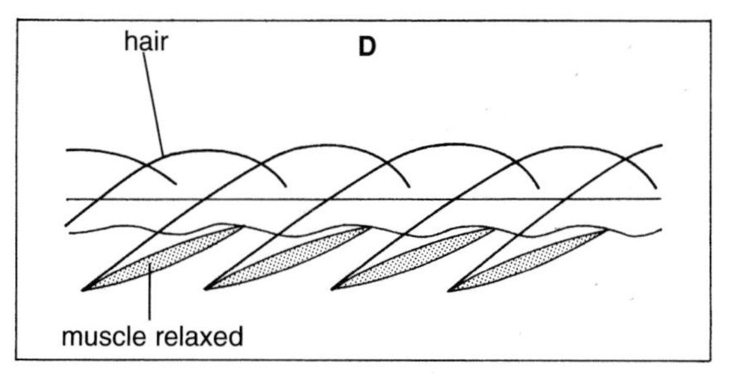

24 The accompanying diagram refers to temperature regulation in the human body. Which situation would be the immediate result of exposure to intense cold?

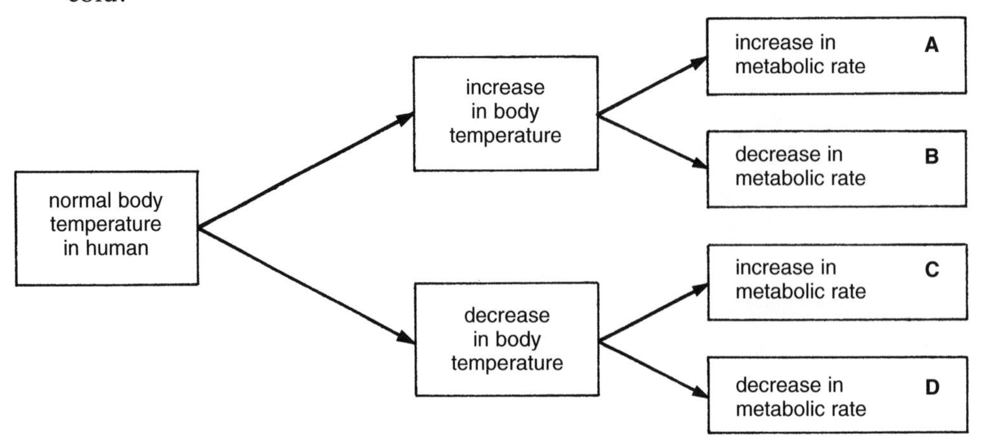

25 The accompanying diagram is incomplete. It shows temperature regulation in the human body under negative feedback control.

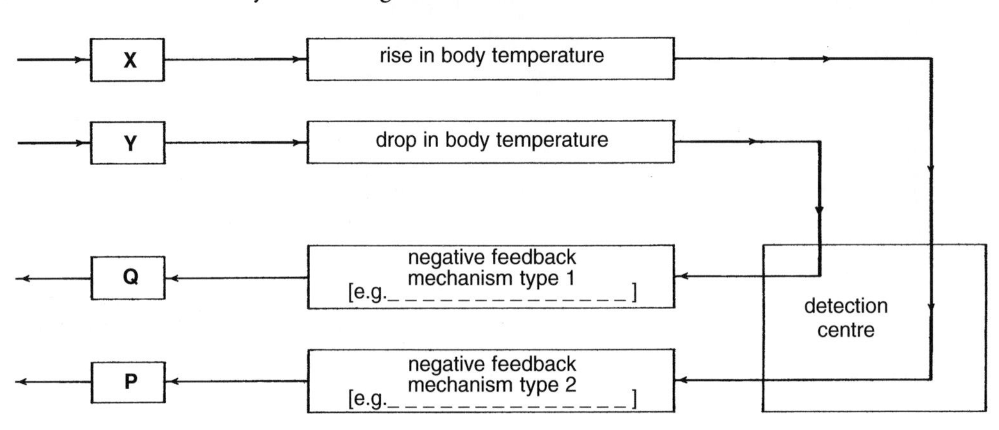

Which line in the following table provides the information needed to complete the diagram?

	Negative feedback mechanism type 1	Lettered arrow to which Q should be joined	Negative feedback mechanism type 2	Lettered arrow to which P should be joined
A	shivering	X	increased sweating	Y
B	shivering	Y	increased sweating	X
C	increased sweating	X	shivering	Y
D	increased sweating	Y	shivering	X

Specimen Examination Paper 1

Items 1 and 2 refer to the accompanying diagram which shows an experiment set up to demonstrate that yeast can respire without oxygen.

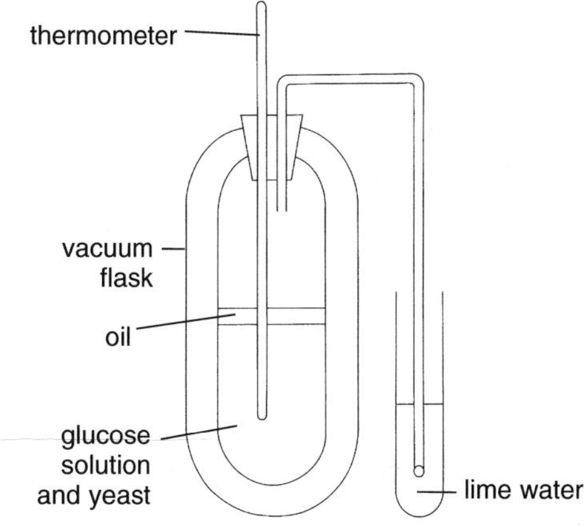

1 The yeast is placed in a vacuum flask in order to

A provide a suitable temperature for growth.
B produce conditions lacking oxygen.
C retain the heat energy produced.
D prevent the evaporation of alcohol.

2 Another name for fermentation in yeast cells is

A anaerobic respiration.
B lactic acid formation.
C aerobic energy release.
D distillation of alcohol.

3 In the situation shown in the accompanying diagram, net movement of water will occur from

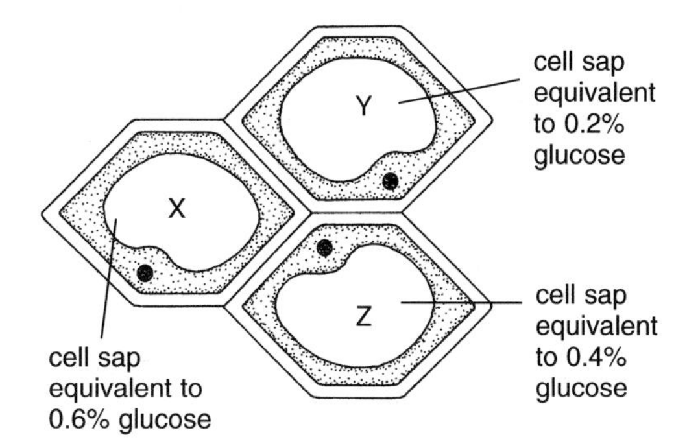

A cell X to cell Y.
B cell Z to cell Y.
C cell X to cell Z.
D cell Y to cell Z.

4 Enzyme molecules are made of

A fat. **B** vitamin.
C protein. **D** carbohydrate.

Items 5 and 6 refer to the accompanying graph which shows the results of an experiment where raw potato containing the enzyme catalase was added to hydrogen peroxide solution at different pH values.

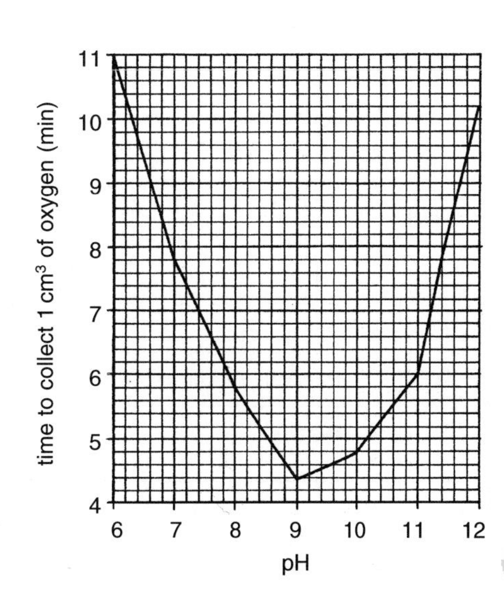

5 The results from this experiment indicate that the optimum pH for the action of the enzyme catalase is

A 6. **B** 9. **C** 11. **D** 12.

6 The time in minutes required to collect 1 cm^3 of oxygen at pH 7.8 would be

A 6.1. **B** 6.2. **C** 7.0. **D** 11.4.

7 The experiment shown in the accompanying diagram was used to compare the amount of heat energy released by two different foodstuffs.

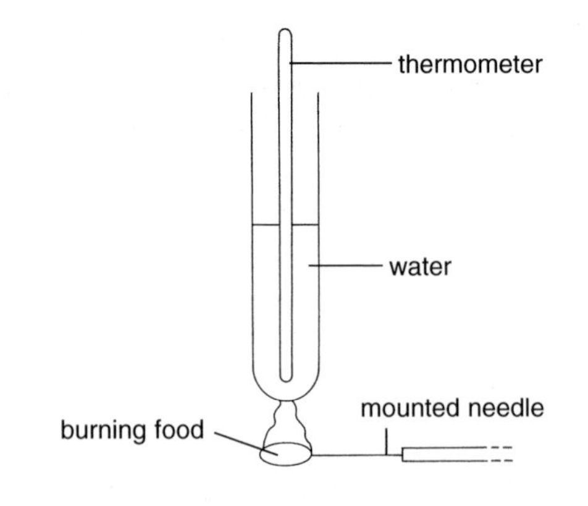

For a valid comparison to be made, the two factors that must be kept constant are

A volume of water and mass of food.
B mass of food and temperature of water.
C temperature of water and type of food.
D type of food and volume of water.

8 During swimming a teenager was found to use energy at a rate of 38 kJ per minute. The energy needed for 10 minutes of this activity could be supplied by consuming 20 g of one of the foods in the accompanying table. Which one?

	Food	**Energy content (kJ/g)**
A	honey	12
B	sucrose	19
C	biscuit	21
D	chocolate	24

9 The accompanying diagram shows a simplified version of the biochemistry of photosynthesis in a chloroplast.

light energy trapped
X
oxygen
hydrogen
Y
hydrogen combined to hydrogen acceptor
Z
glucose

Which line in the table identifies the chemical substances that should have been named in boxes X, Y and Z?

	X	**Y**	**Z**
A	carbon dioxide	hydrogen acceptor	water
B	water	carbon dioxide	hydrogen acceptor
C	hydrogen acceptor	water	carbon dioxide
D	water	hydrogen acceptor	carbon dioxide

10 Which of the following defines the term *ecosystem*?

A the place where an organism lives
B a group of organisms of the one species
C all of the organisms that live together in a region
D a natural biological unit made of living and non-living parts

11 Three food chains are summarised in the following table.

	Food chain		
	1	**2**	**3**
tertiary consumer	gull	flea	sparrow
secondary consumer	crab	barn owl	ladybird
primary consumer	mussel	field mouse	greenfly
producer	phytoplankton	wheat plant	rose bush

The pyramid of numbers shown in the accompanying diagram could be used to correctly represent

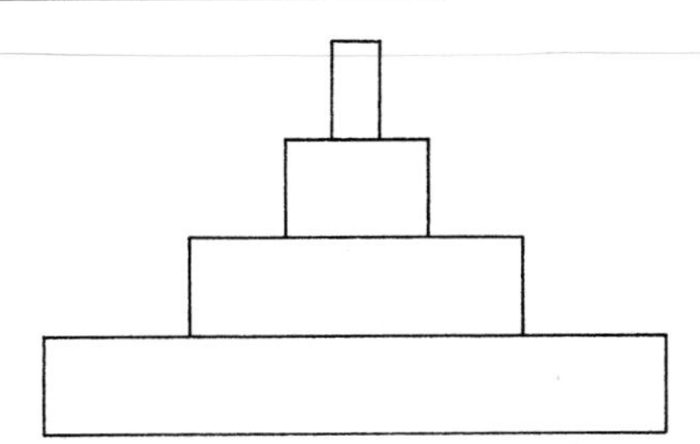

A food chain 1 only.
B food chains 1 and 3 only.
C food chains 2 and 3 only.
D food chains 1, 2 and 3.

12 Which of the following factors BOTH cause habitat destruction?

A desertification and conservation
B conservation and deforestation
C deforestation and pollution
D pollution and biodiversity

13 The accompanying diagram represents a choice chamber.

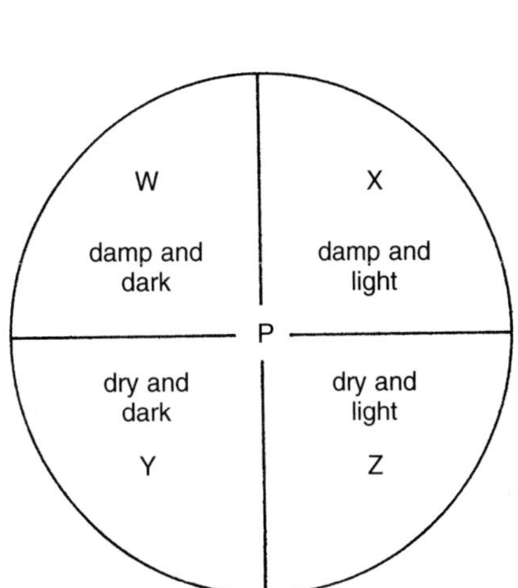

If 25 woodlice were released at point P, the most likely distribution after 10 minutes would be

A 10 in W, 3 in X, 9 in Y, 3 in Z.
B 6 in W, 6 in X, 6 in Y, 7 in Z.
C 19 in W, 3 in X, 2 in Y, 1 in Z.
D 2 in W, 1 in X, 19 in Y, 3 in Z.

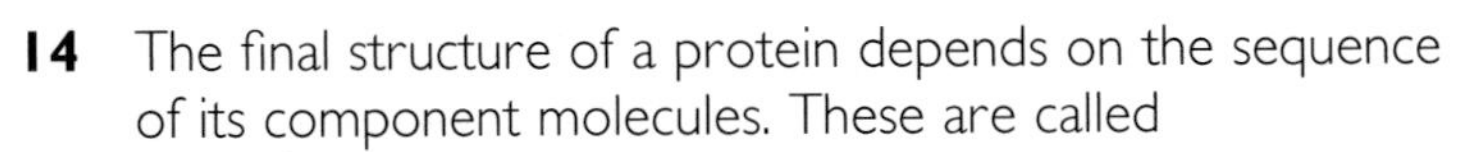

14 The final structure of a protein depends on the sequence of its component molecules. These are called

A amino acids. **B** bases.
C nucleic acids. **D** enzymes.

15 Which of the following does NOT contain a double set of matching chromosomes?

A brain cell **B** cheek cell
C sperm cell **D** egg mother cell

16 Albinism is determined by a recessive allele (a). A woman with normal skin pigmentation but whose father was an albino, marries an albino man.

What is the chance of each of their children being albino?

A 1 in 1 **B** 1 in 2
C 1 in 3 **D** 1 in 4

17 In guinea pigs, long hair is recessive to short hair. If a large group of heterozygous females are crossed with a large group of heterozygous males, the percentage of their offspring with long hair will be approximately

A 25. **B** 50. **C** 75. **D** 100.

18 Which of the following statements refers CORRECTLY to selective breeding of crop plants?

A The plant's genotype is altered directly by manipulating its chromosomal material.
B The new variety formed can make products previously only made by another species.
C The gene for the useful characteristic is transferred from the plant to a bacterium.
D The production of an improved variety involves many years of repeatedly choosing the best plants.

19 The accompanying diagram shows the human alimentary canal. In which two regions does digestion take place?

A 1 and 2
B 1 and 3
C 2 and 3
D 3 and 4

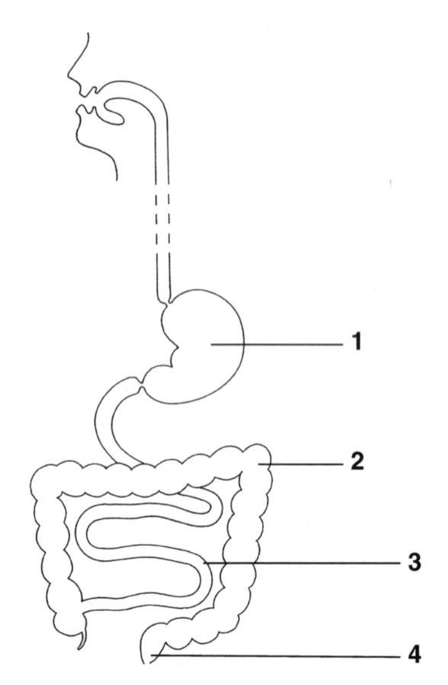

20 Iron is an essential constituent of

A ATP. **B** bone.
C DNA. **D** haemoglobin.

21 The accompanying diagram shows part of a kidney nephron and its blood supply.

Which letter indicates the glomerulus?

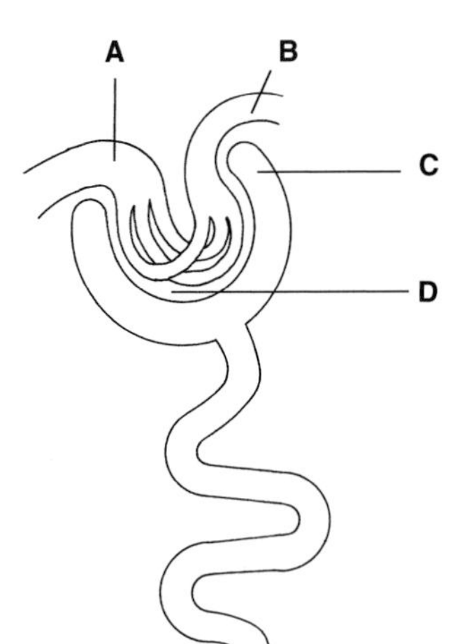

22 A woman runs a marathon, sweats profusely and drinks little fluid. Which line in the table CORRECTLY summarises the events that result from this behaviour?

	ADH production	Water reabsorption	Urine output
A	↑	↑	↓
B	↑	↓	↓
C	↓	↑	↑
D	↓	↓	↑

↑ = increase ↓ = decrease

23 The accompanying diagram shows the human heart. Vessel X carries

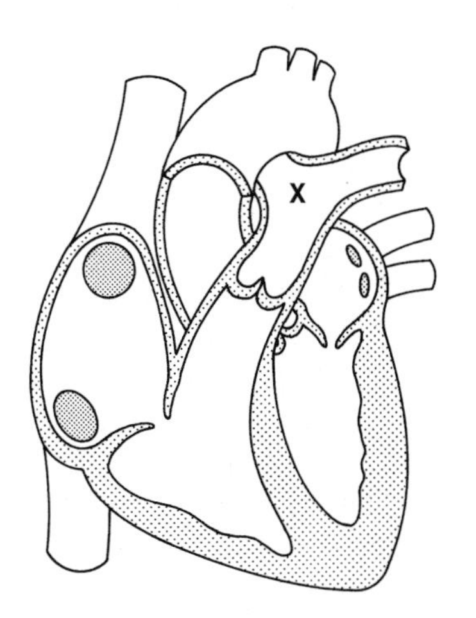

A oxygenated blood from the heart.
B deoxygenated blood to the lungs.
C oxygenated blood from the lungs.
D deoxygenated blood to the heart.

24 Which of the following statements is CORRECT?

A An antibody stimulates white blood cells to make antigens.
B An antibody is always composed of viral protein.
C An antigen stimulates white blood cells to make antibodies.
D An antigen is always composed of viral protein.

25 The following list gives five stages that occur during a reflex action.

1 nerve impulse transmitted through sensory neurone
2 response made by effector
3 nerve impulse transmitted through relay neurone
4 nerve impulse transmitted through motor neurone
5 stimulus detected by sensory receptor

Which of the following is the correct sequence of events?

A 1, 5, 3, 2, 4 **B** 5, 1, 3, 4, 2
C 1, 5, 4, 3, 2 **D** 5, 1, 3, 2, 4

Specimen Examination Paper 2

1 The accompanying diagram shows *Euglena*. This unicellular organism could be classified as a plant due to the presence of a

A flagellum. **B** chloroplast.
C cell membrane. **D** contractile vacuole.

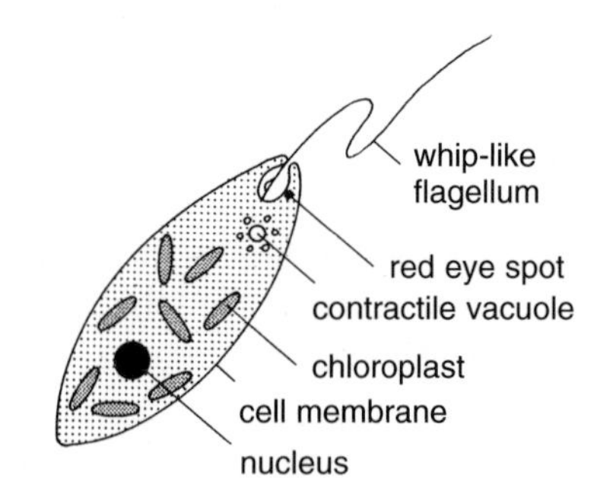

Items 2 and 3 refer to the following information. In an experiment, groups of potato discs were weighed and then each group was immersed in one of a series of sucrose solutions. After two hours each group was reweighed and its percentage gain or loss in weight was calculated.
The accompanying graph shows the results plotted as points.

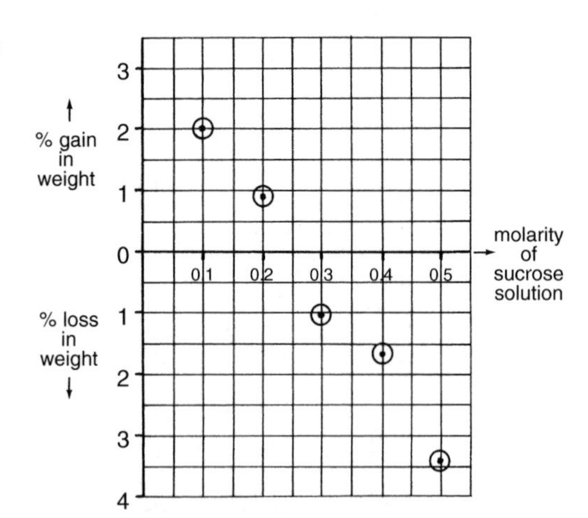

2 From these results, it can be concluded that the water concentration of the potato cell sap is approximately equivalent to that of a sucrose solution of molarity

A 0.10. **B** 0.25.
C 0.35. **D** 0.50.

3 The table lists aspects of good practice normally carried out during this investigation. Which line CORRECTLY pairs a practice with the reason for carrying it out?

	Good practice	**Reason**
A	all factors kept equal except the one under investigation	to obtain a reliable set of results
B	class results pooled and averages calculated	to remove excess liquid which would invalidate the results
C	discs blotted on paper towels	to ensure that no second variable factor was included in the investigation
D	results converted to percentages	to standardise the results because the initial masses of discs may not have been identical

4 The following table gives the results from an experiment involving the digestion of a food by an enzyme.

Temperature (°C)	5	15	25	35	45	55
Rate of digestion (mg/h)	3	8	17	21	18	1

Which of the accompanying graphs represents this information accurately?

A

B

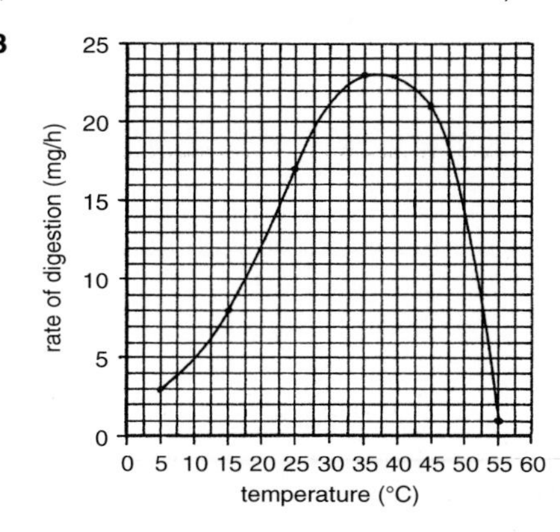

C

D

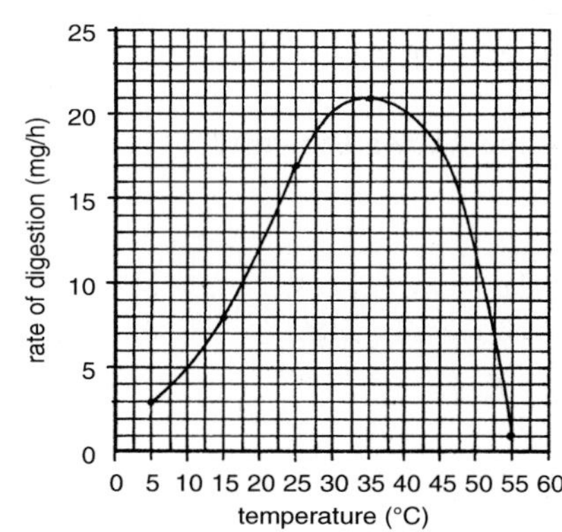

5 Which of the following terms is the name given to a stage of the biochemical pathway that is common to aerobic and anaerobic respiration?

A glycogen **B** glycolysis
C lactic acid **D** pyruvic acid

6 Glucose contains 17 kJ/g. Which of these activities, when carried out for 20 minutes, uses the same amount of energy as is contained in 50 g of glucose?

	Activity	**Energy used (kJ/min)**
A	playing football	37.5
B	walking upstairs	38.5
C	rowing	40.5
D	running	42.5

7 Diffusion is important to a green leaf cell during photosynthesis because it is how

A carbon dioxide, a useful substance, enters and oxygen, a waste product, leaves.
B oxygen, a useful substance, enters and carbon dioxide, a waste product, leaves.
C carbon dioxide, a waste product, enters and oxygen, a useful substance, leaves.
D oxygen, a waste product, enters and carbon dioxide, a useful substance, leaves.

8 Which line in the following table refers CORRECTLY to photosynthesis?

	Light-dependent stage	Temperature-dependent stage
A	photolysis	carbon fixation
B	carbon fixation	photolysis
C	photolysis	respiration
D	chloroplast	carbon fixation

9 The accompanying diagram shows a pyramid of numbers of organisms in an ecosystem.

Which level represents the primary consumers?

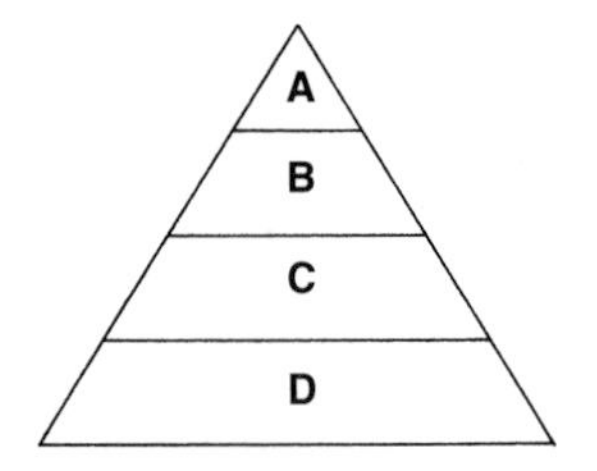

10 The accompanying table shows the concentration of a non-biodegradable pesticide residue in the tissues of several organisms in a food chain and in the water of their ecosystem.

The concentration of pesticide increased by a factor of 9 between

A herbivorous fish and carnivorous fish.
B carnivorous fish and fish-eating bird.
C plankton and herbivorous fish.
D water and plankton.

Substance or organism analysed	Concentration of pesticide (ppm)
water	0.00005
plankton	0.04
herbivorous fish	0.23
carnivorous fish	2.07
fish-eating bird	6.00

11 Which line in the following table is CORRECT?

	Organism	Habitat	Niche
A	toadstool	woodland floor	producer that feeds on dead oak leaves and recycles mineral salts round ecosystem
B	heather	moorland	producer adapted to life in exposed environment and eaten by consumers such as red grouse
C	pond snail	fresh water	carnivorous vertebrate which scavenges dead remains in pond mud
D	shore crab	seashore	herbivorous invertebrate which actively seeks prey when tide is in

12 The accompanying diagram shows the female reproductive organ of a flower.

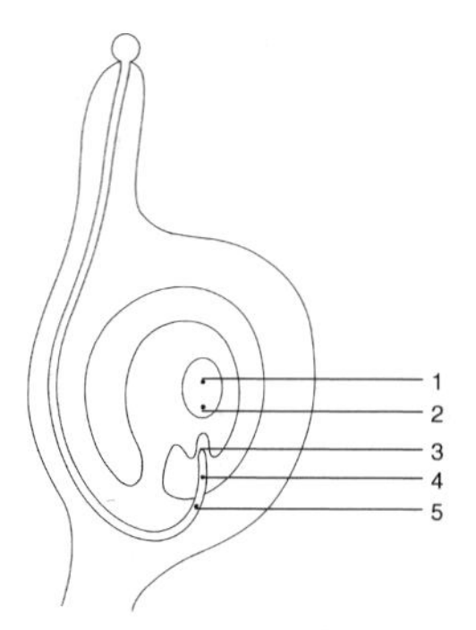

Which structures fuse to form a zygote at fertilisation?

A 1 and 2 **B** 2 and 3
C 2 and 4 **D** 4 and 5

13 The first of the accompanying diagrams shows the chromosomes present in the nuclei of various types of cell from the fruit fly, *Drosophila*.

Which part of the second diagram represents the set of chromosomes that would be found in a zygote formed as a result of sperm type 2 fertilising a normal egg?

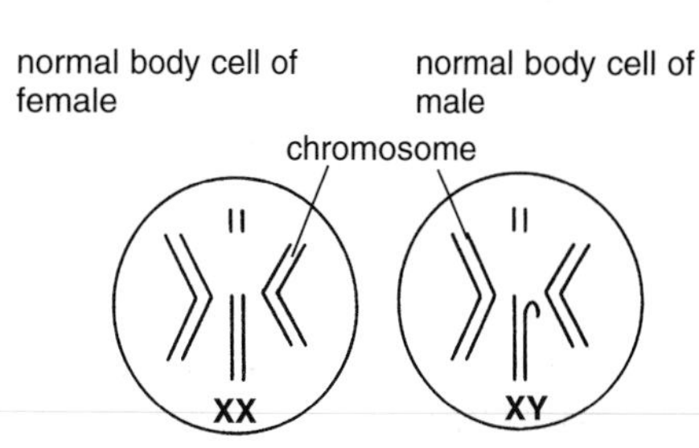

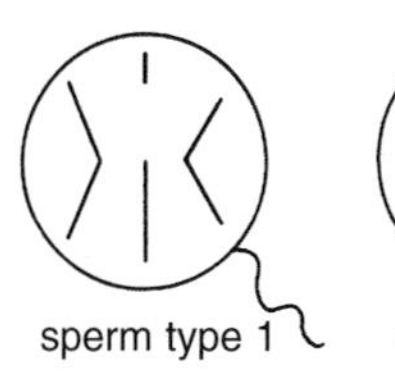

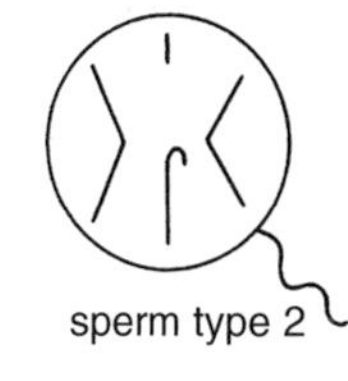

A

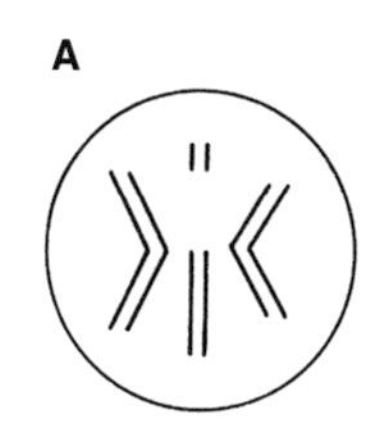

B

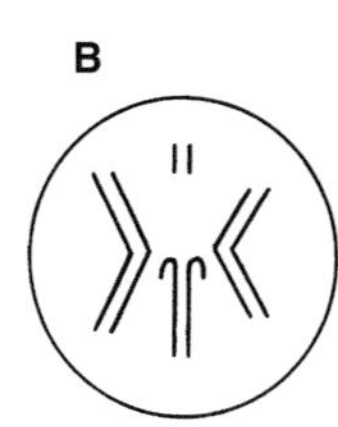

C

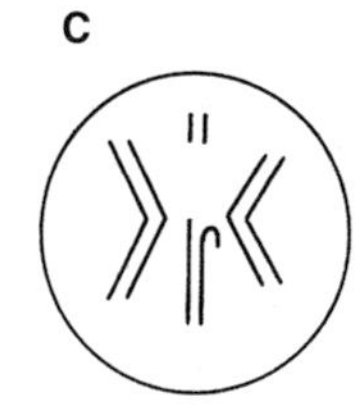

D

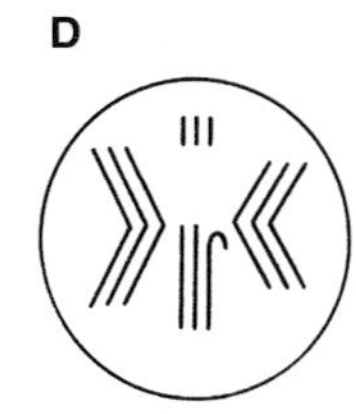

14 In pea plants, the allele for tall height (T) is dominant to the allele for short height (t). In which of the following is the phenotypic ratio of the offspring correct for the cross indicated?

	Cross	**Phenotypic ratio of offspring**
A	Tt × tt	all tall
B	Tt × Tt	1 tall : 1 short
C	Tt × Tt	1 tall : 3 short
D	Tt × tt	1 tall : 1 short

15 A cell which carries two different alleles of a gene is said to be

A heterozygous. **B** homozygous.
C dominant. **D** recessive.

16 Genetic engineering is the process by which

A a piece of DNA is removed from one species and inserted into another.
B a piece of DNA is removed from a diseased organism and discarded.
C a reprogrammed species is crossed with a species possessing different DNA.
D a sample of sperm containing paternal DNA is kept frozen ina sperm bank until required.

17 The accompanying diagram charts the progress of four types of food along the human alimentary canal. Which type of food undergoes digestion in the mouth and small intestine?

type of food	in mouth	in stomach	in small intestine
A			
B			
C			
D			

KEY [shaded] = undigested food [blank] = enzyme-digested food

18 The accompanying diagram shows a molecule of fat. Structure X is a molecule of

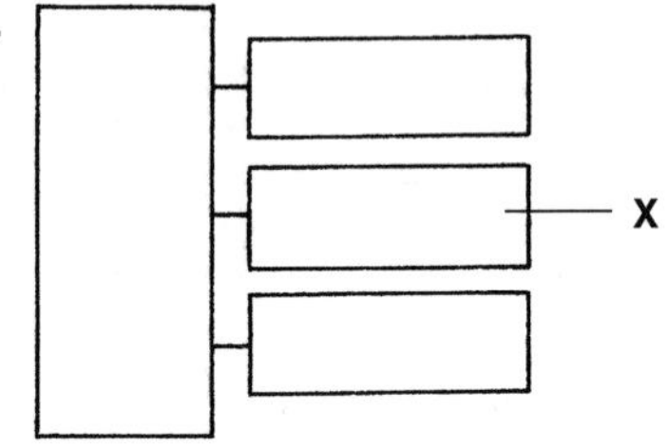

A glycogen. **B** glycerol.
C fatty acid. **D** amino acid.

19 The water concentration of a salt water fish's body is maintained at the optimum level by

A a large volume of dilute urine being produced.
B a large volume of the surrounding sea water being drunk.
C mineral salts being absorbed by specialised cells in the gills.
D blood passing through the kidneys being filtered at a high rate.

20 The accompanying table shows the results of an investigation into the composition of three types of renal fluid in an adult male.

Type of renal fluid	Concentration of substance in renal fluid (g/100 ml)				
	Amino acids	Salts	Protein	Urea	Glucose
plasma	0.05	0.72	8.00	0.03	0.10
glomerular filtrate	0.05	0.72	0.00	0.03	0.10
urine	0.00	1.44	0.00	2.10	0.00

The two substances that are completely reabsorbed back into the bloodstream are

A urea and salts.
B glucose and protein.
C amino acids and glucose.
D protein and amino acids.

21 The type of blood vessel shown in the accompanying diagram is

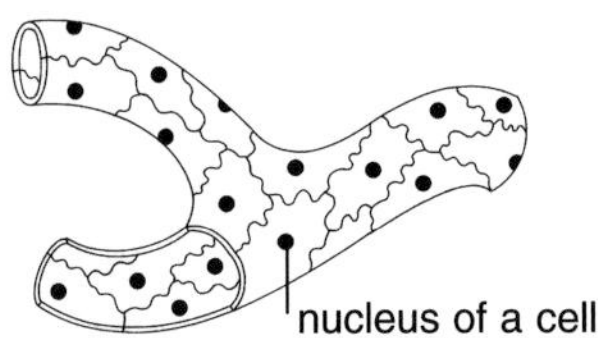

A a vein. **B** an artery.
C a capillary. **D** an arteriole.

22 In the accompanying diagram, a capillary is in close contact with structure Y. The relative concentration (in units) of carbon dioxide (CO_2) and oxygen (O_2) are given at three different sites. Which line in the table is CORRECT?

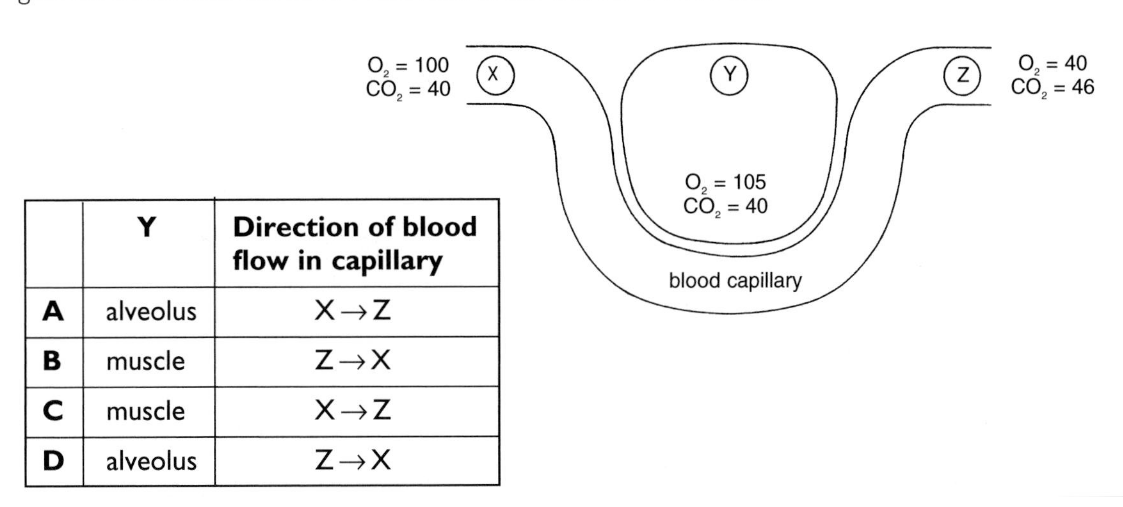

	Y	Direction of blood flow in capillary
A	alveolus	X→Z
B	muscle	Z→X
C	muscle	X→Z
D	alveolus	Z→X

23 Haemoglobin is a respiratory pigment which

A retains oxygen at low oxygen levels in the lungs.
B releases oxygen at high oxygen in the lungs.
C retains oxygen at high oxygen levels in the tissues.
D releases oxygen at low oxygen levels in the tissues.

24 Which column in the following table describes a reflex action?

	A	B	C	D
rapid	✓	✗	✓	✓
slow	✗	✓	✗	✗
protective	✓	✓	✓	✗
non-protective	✗	✗	✗	✓
voluntary	✗	✗	✓	✗
involuntary	✓	✓	✗	✓

25 Which of the following parts of the human body does NOT act as a temperature-regulating effector?

A sweat gland
B pituitary gland
C skeletal muscle
D hair erector muscle

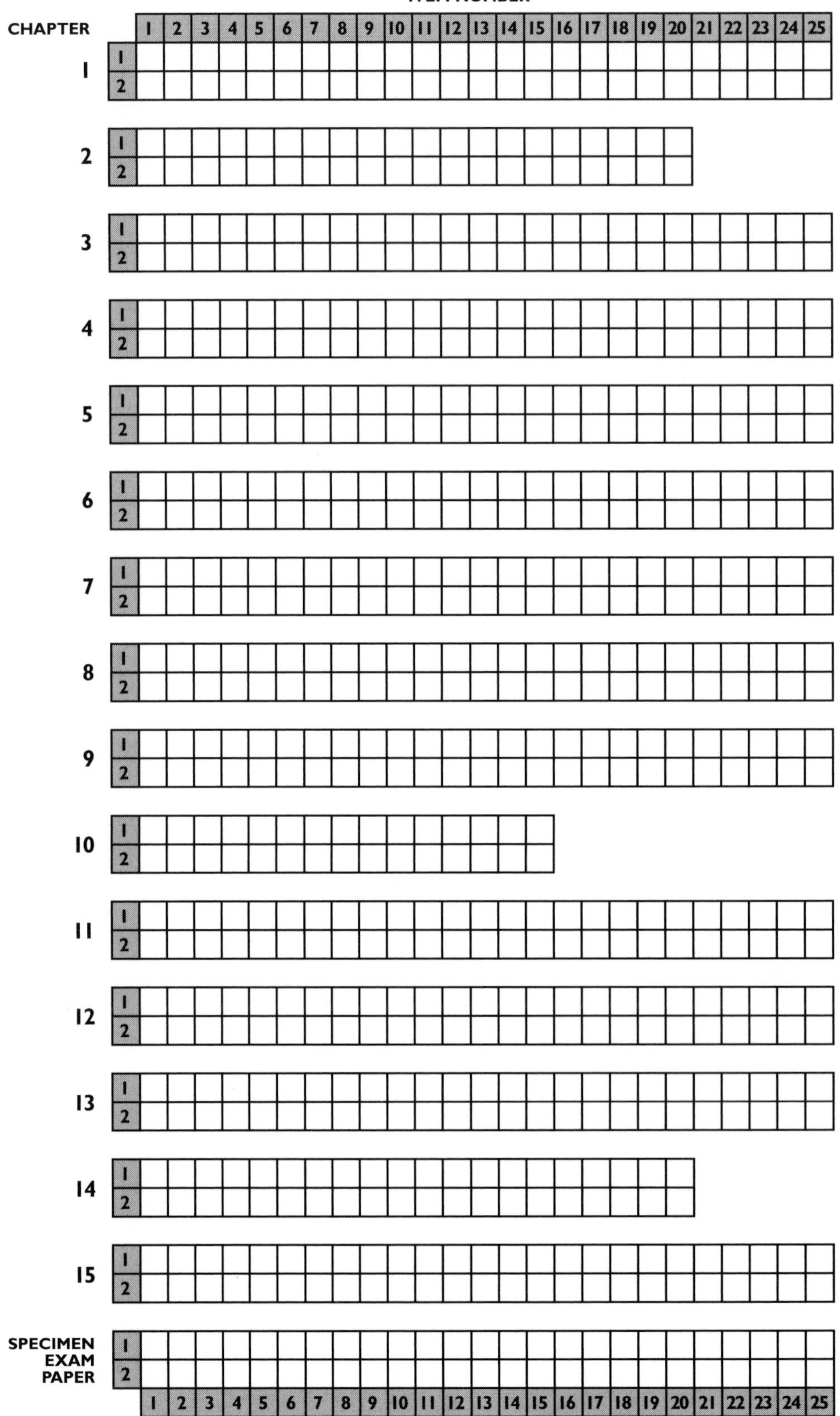

ITEM NUMBER

CHAPTER		1	2	3	4	5	6	7	8	9	10	11	12	13	14	15	16	17	18	19	20	21	22	23	24	25
1	1																									
	2																									
2	1																									
	2																									
3	1																									
	2																									
4	1																									
	2																									
5	1																									
	2																									
6	1																									
	2																									
7	1																									
	2																									
8	1																									
	2																									
9	1																									
	2																									
10	1																									
	2																									
11	1																									
	2																									
12	1																									
	2																									
13	1																									
	2																									
14	1																									
	2																									
15	1																									
	2																									
SPECIMEN EXAM PAPER	1																									
	2																									
		1	2	3	4	5	6	7	8	9	10	11	12	13	14	15	16	17	18	19	20	21	22	23	24	25